AF522252

GENOMIC EVOLUTION

ENCYCLOPAEDIA OF GENE EVOLUTION

Vol. IV

GENOMIC EVOLUTION

By

Dr. M. Prakash

DISCOVERY PUBLISHING HOUSE
NEW DELHI-110002

First Published-2006

ISBN 81-8356-169-1 (Set)

Published by

DISCOVERY PUBLISHING HOUSE
4831/24, Ansari Road, Prahlad Street,
Darya Ganj, New Delhi-110002 (India)
Phone: 23279245 • Fax: 91-11-23253475
E-mail:dphtemp@indiatimes.com

Printed at:
Amit Enterprises
Delhi-53

PREFACE

The present title **Encyclopaedia of Gene Evolution** encompasses diverse areas of study from the origin of life and is subsequent record to the molecular and population concepts that explain these events. All biological phenomena derive from evolutionary relationships and past interactions. As the great evolutionary geneticist Theodosius Dobzhansky remarked, "Nothing in biology makes sense except in the light of evolution" yet unifying all biology under an evolutionary theme is still difficult, although the explosive increase in molecular, organismic, and populations information makes the realization of this goal more possible now than ever before.

The aim in writing this text was to bring together the highly dispersed literature on gene structure, function and expression, to integrate this with our emerging knowledge of chromosome and genome structure, and to draw comparison both within and between paralogous and orthologous gene sequences in order to establish the nature of the mutational mechanisms responsible for their evolutionary divergence. Such an approach, encompassing a large number of human genes and their extended families, is essential for any attempt to discern underlying evolutionary principles.

The purpose of this book is to bring together some prevailing knowledge and ideas about evolution, to provide an informed framework of together for under graduates, post-graduate and those of related fields.

Since evolution is the broadest of biological fields, covering the greatest range of disciplines, even the brief survey of evolution offered here has errors and ambiguities. To the extent that this book has been spared many such failings the author owes thanks to many reviewers we commented on one or more sections of this text.

There can be no claim to originality except in the manner of treatment and much of the information has been obtained from the books and scientific journals available in the different libraries.

The author expresses his thanks to his friends and colleagues whose continue inspirations have initiated him to bring out this book.

The author expresses his gratitude to Mr. Wasan and staff of M/s Discovery Publishing House for their whole hearted co-operation in the publication of this book.

Author

CONTENTS

1

INTRODUCTION

The haploid human genome comprises 23 chromosomes containing between them some 3.2 × 10^9 bp of DNA. The bulk of the genome comprises DNA sequence of varying degrees of repetitivity whilst about 10% represents unique (single copy) sequence containing perhaps 80000 genes. The repetitive portion contains a mixture of DNA sequence elements which may have a structural or regulatory role or may be merely 'junk DNA' without obvious function. Human genes contain within them all the genetic information necessary to specify the encoded proteins but they also contain further information, in the form of a large number of different DNA sequence motifs that serve to control mRNA expression, splicing, transport, and stability. This chapter constitutes a short introduction to the structure, function and expression of the human genome together with a brief description of the types of mutational lesion in human genes that have been found to be responsible for inherited disease.

CHROMOSOME STRUCTURE AND FUNCTION

Human chromosomes contain DNA in a highly coiled and condensed form, organized and packaged by structures known as *nucleosomes*. Chains of nucleosomes comprise a '10 nm fibre' and this is coiled to form the '30 nm fibre' which is in turn further coiled to form *chromatin*. The highest degree of condensation is found in transcriptionally inactive regions, those regions in which mRNA synthesis from specific genes is turned off. To allow transcriptional activation of a gene (i.e. to allow the enzyme RNA polymerase to initiate mRNA synthesis), chromatin must be uncoiled, a process which occurs in at least three stages: the unfolding of large chromosomal domains (25-100 kb), the remodeling of the chromatin structure of gene regulatory regions and the alteration of nucleosome structure in transcribed regions. Unfolding reveals binding sites on the chromosomal DNA for activator proteins. Once bound, these proteins alter nucleosome positioning thus provides an accessible template for the assembly of the transcriptional initiation complex.

Centromeres

The centromere is essential for normal disjunction of the chromosomes following cellular division at meiosis and mitosis. Centromeric DNA consists of arrays of tandemly repeated DNA sequences which have undergone homogenization by repeated genetic exchange. Alphoid satellite DNA is the only satellite DNA family known to be present in the centromeric regions of all human chromosomes. The basic 169 and 172 bp repeats of this primate specific satellite DNA comprise the bulk of human centromeric DNA and contain a 17 bp binding site for the centromere-specific protein CENP-B. This binding site motif is present at the centromeres in the chromosomes of all the anthropoid apes but is absent from the genomes of Old and New World monkeys and prosimians. It is also absent from human Y chromosomal alphoid satellite DNA.

Human alphoid satellite DNA may be divided into subsets which are largely chromosome-specific. They may however be groped into four supra-chromosomal families each characterized by a specific monomer type. Homologies exist between alphoid satellite DNAs from different ape species but these are not associated with homologous chromosomes. Although alphoid satellite repeats may have evolved from a common ancestral repeat monomer, they have also been subject to concerted evolution between homologous chromosomes within a given species.

Using a combination of oligonucleotide primer extension and immunocytochemistry, Mitchell et al. (1992) showed that the alphoid repeats were closely associated with the *kinetochore* (the structural element on the chromosome that binds to the mitotic spindle). The presence of (AATGG)n (CCATT)n repeats in the centromeric region suggests that stem-loop structures might form which could serve as specific recognition sites for kinetochore function. Alphoid satellite DNA sequences are not however restricted to centromeric regions.

Telomeres

Telomeres allow the end of chromosomal DNA to be replicated completely without the loss of bases at the termini. They are the sites at which the pairing of homologous chromosomes is initiated and in humans contain long arrays (averaging about 10-15 kb) of minisatellite DNA comprising tandem hexanucleotide repeats, most frequently TTAGGG. Other telomeric hexanucleotide repeats (e.g. TTGGGG, TGAGGG) are also known. These sequences and their variants are tandemly repeated but are nonrandomly distributed and polymorphic in terms of their location. Chimpanzees also possess the TTAGGG telomeric repeat but differ from humans in terms of their subterminal satellite sequences.

The simple sequence of telomeres is synthesized by the ribonucleoprotein polymerase, *telomerase*, and is thought to protect the ends of chromosomes from degradation during DNA replication. Telomere length decreases with age and number of cell divisions. Some chromosomes (e.g., 9p, 12p, 14p, 17p, 21p, and 9p) have telomeres shorter than the average whereas others (e.g. 4q, 5p, 18q and Xp) have telomeres which are longer. Telomere length polymorphism are apparent for some chromosomes, for example 11p.

Sites of Recombination

Chiasmata represent the cytological evidence for recombination. At meiosis, each pair of homologous chromosomes possesses at least one chiasma and the number of chiasma per pair is proportional to the size of the chromosome. Chiasma frequency is a function both of distance from the centromere and chromosome band 'flavor': the G+C rich T bands exhibit a six-fold higher chiasma frequently than G bands. Since G+C content is in general positively correlated to chiasma frequency, this could explain why gene density is higher in T bands.

Recombination is considerably higher in human females than in males, as evidenced by the average distance between two markers in the female genetic map being 85% longer than in males. Chiasma frequency is a function of distance from the centromere. Recombination therefore tends to increase towards the telomeres, the distal 15% of chromosomes containing 40% of the chiasmata. Finally, it may also be pertinent to consider that intrachromosomal homologous recombination may be enhanced by transcription in mammalian cells.

Obligatory recombination occurs during male meiosis within the pseudoautosomal region, a 2.6 Mb stretch of homologous sequence at the tip of the short arms of the X and Y chromosomes. Genes in this region escape X-inactivation and the boundaries of this region appear to have been conserved evolutionarily between Old World monkeys and human.

Regions of sex-specific hypo- and hyper-recombination have been reported in a study which compared genetic and physical maps of human chromosome 19. Other recombination hotspots have been characterized in specific human genes including those encoding the T-cell receptor beta chain and HLA-associated ATP transporter 2 loci. Several different types of DNA sequence have been proposed to be recombinational hotspots in the genomes of mice and men. $(CAGA)_6$ and $(CAGG)_{7\text{-}9}$ represent hotspots of recombination in the murine MHC gene cluster. Other sequences thought to promote recombinational instability are alphoid repeats, a marine transposon-like element, Z-DNA ('left-handed DNA') and minisatellite sequences. Recombinational breakpoints have also been found to be associated with topoisomerase I cleavage sites in the rat genome. The majority of these cleavage sites contain the sequences CTT and GTT. It may therefore be that the process of nonhomologous recombination is mediated by topoisomerase I.

Gene Distribution and Density

Several thousand genes have now been mapped to within single chromosome bands. Some 80% map to the G-C rich R bands whilst 20% map to G bands. A similar distribution is apparent for CpG islands: 86% are located in R bands. 'Housekeeping' genes are strictly confined to the R bands together with about half of the tissue-specific genes whereas the remainder of the tissue-specific genes are present in the G-bands. One the four recognized types or 'flavours' of R band, known as T bands, are often found at telomeres, exhibit the highest G +C content and contain between 58% and 68% of R band genes as well as the majority of CpG islands.

Chromosomes 13 and 18 appear to possess a relatively low gene density and chromosome 19 a relatively high density as evidenced by the chromosomal assignment of some 320 cDNAs derived from a human brain cDNA library. Interestingly, DNA excision repair may be preferentially directed toward regions of high gene density, a reflection perhaps of the preferential repair of actively transcribed gene sequences.

Isochores

The human genome is a mosaic of large (>300 kb) DNA segments or isochores that are compositionally homogeneous and which can be subdivided into a small number of families characterized by different degrees of GC-richness (30-60%). Five families have been identified: L1 and L2 which are GC-poor and comprise 62% of the genome and H1, H2, and H3 which are GC-rich and represent 22%, 9% and 3%

of the genome respectively. Gene concentration varies between isochores: 34% of human genes are located in L1 and L2 isochores, 38% in the H1 and H2 isochores, and 28% in the H3 isochores. The banding pattern of chromosomes reflects the isochore organization: thus, the G bands are formed by L1 and L2 isochores whilst the T bands are formed by the H2 and H3 families. A recent study of two primate globin pseudogenes which reside in different isochore compartments has provided evidence that isochores have arises as a result of mutational bias rather than from the action of selection.

Matrix Attachment Regions

Chromatin is attached to the nuclear matrix or scaffold at specific sites known as *matrix* or scaffold attachment regions (MARs/SARs). The organization of chromatin with respect to the nuclear scaffold is thought to determine chromosome architecture in terms of its functional domains; this in turn influences gene activity. Indeed, MARs may function so as to place genes at the nuclear scaffold in order to facilitate their transcription. MARs do not appear to share extensive sequence homology but often comprise 200 bp of AT-rich DNA, for example the sequence of AATATTTTT in the murine immunoglobulin k gene locus. A number of MAR-binding proteins are known to bind to Mars including the attachment region-binding protein, ARBP and histone H1.

MARs appear to be preferentially associated with topoisomerase II cleavage sites and share sequence homology with binding sites for homeobox proteins. Topoisomerase II plays a role in the segregation of daughter chromosomes after DNA replication and also in chromosome condensation; it binds preferentially to MARs. Vertebrate topoisomerase II cleavage sites also occur in association with MARs and manifest a consensus sequence, A/G N T/C N N C N N G T/ C N G G/T T N T/C N T/C.

MARs also appear to be preferentially associated with enhancer-type elements. Indeed, MARs stimulate heterologous gene expression in reporter gene experiments. A cis-acting regulatory element 3' to the human $^{A}\gamma$-globin (HBG1) gene, known to be associated with the nuclear matrix, has been shown to bind specifically to an AT-rich binding protein (SATB1) that binds to MARs.

Origins of DNA Replication

Chromosomal DNA replication initiates at specific points (*origins*) and proceeds outward bidirectionally from specific loci. Although a number of putative origins of replication have been identified in mammalian species data are still sparse. One of the best characterized is that found in human between the δ-(HBD) and β-globin (HBB) genes on chromosome 11. This replication origin is bidirectional and functional regardless of the transcriptional state of the β-globin gene. From the study of six putative origins of replication (including one in the human c-myc oncogene), Dobbs et al. (1994) claimed to have derived a consensus sequence, albeit a fairly redundant one: A/T A A/T T T/A/G/T A/G/T A/T A/T A/T A/G/T A/C/T A/T G A/T A/C/T A/C A A/T T T. However, there are probably several different classes of replication origins which possess different sequence characteristics.

Replication origins are often associated with CpG-rich regions and may sometimes be located in the vicinity of matrix attachment regions. In the human genome, the units of DNA replication range in size from 50 kb to 600 kb and are often clustered. For instance, there are at least six such *replicons* within the human dystrophin (DMX; Xp21) gene. The gene-rich R bands replicate early in S phase whilst the G bands replicate late. Housekeeping genes invariably replicate early whilst tissue-specific genes can be early or late replicating (some replicate earlier when transcriptionally active). Nontranscribed genes on the X chromosome also replicate late.

Gene Organization and Transcriptional Regulation

Gene Structure and Regulation

The coding portion of the human genome, roughly 5% of the total DNA complement, probably contains some 70,000 different gene sequences. Thus, the smallest human chromosome, 21, may contain as many as 2000 genes.

It is now known that the most genes in higher organisms are not contiguous but rather are a complex mosaic or protein coding (*exon*) and intervening non-coding (*intron*) sequences. The exons represent that portion of the gene which encodes the amino acid sequence of the protein product plus 5' and 3' noncoding regions. Initially, both exons and introns are *transcribed* into mRNA but the intronic portion is ultimately removed during mRNA maturation by a process known as *splicing*. The mature of mRNA is then *translated* into the amino acid sequence of a protein on the ribosome. Although the central dogma of molecular biology was therefore once summarized as 'DNA makes RNA makes protein,' the reverse flow of genetic information is also possible by *reverse transcription* of RNA into DNA (copy DNA or cDNA).

Each individual gene differs not only with respect to its DNA sequence specifying the amino acid sequence of the protein it encodes,

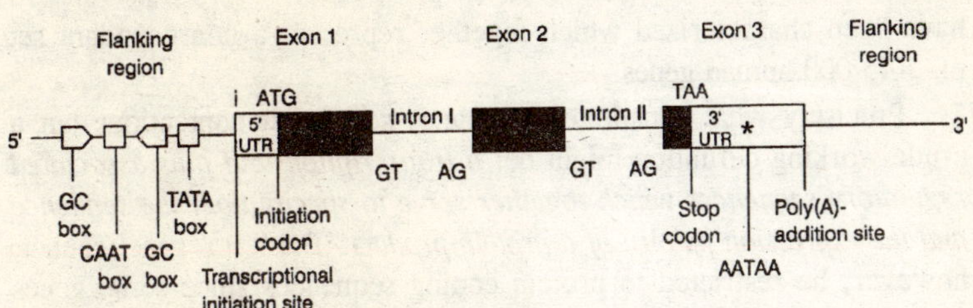

Fig. 1.1. Schematic structure of an archetypal human protein-coding gene.

but also with respect to its structure. A few human genes are devoid of introns (e.g. thrombomodulin THBD) which spans 3.7 kb) whereas others may possess a considerable number, for example 79 in the 2.4 Mb dystrophin (DMD) gene and as many as 118 in the α1(VII) collagen (COL7A1) gene. Introns maybe classified according to whether they interrupt the reading frame of the encoded protein. Thus phase 0 denotes that the intron lies between two codons, phase 1 between the first and second nucleotides of a codon, and phase 2 between the second and third nucleotides of a codon.

Some introns may be huge as in the case of the first intron of the human COL5A1 gene (~600 kb). The average length of a vertebrate intron has been estimated to be ~620 bp but introns separating exons preceding the coding ones are often rather longer with an average length of >1800 bp. This suggests that evolution may sometimes have had to trawl quite far upstream of a gene to recruit appropriate DNA sequence motifs to act as promoter/regulatory elements within the 5' untranslated region. The average length of an internal exon is ~140 bp but this average figure conceals some very large exons, for example in the human factor VIII (F8C; Xq28) [3106 bp], apolipoprotein B (APOB; 2p23-p24) [7572 bp] and mucin 5B (MUC5B; 11p15.5) [10,690 bp] genes.

Some human genes encode very large proteins, for example the cardiac titin (TTN; 2q31) cDNA is 82 kb in length and predicts a 27,000 amino acid protein with a molecular weight of nearly 3000 kDa. Large mRNAs are also generated from the dystrophin (DMD; Xp21.2-2.13 14 kb), apolipoprotein B (APOB; 2p23-24; 14 kb), and mucin (membrane-associated glycoprotein) genes MUC2, MUC5AC, MUC5B, MUC6 (11p15.5; 15-18 kb) MUC3 (7q22; 17 kb), and MUC4 (3q24; 18-24 kb).

Only ~10% of human genes encode proteins of known function. The sequences of many of these genes can be found in GenBank. In addition, hundreds of thousands of 'expressed sequence tags' (ESTs) have been characterized which together represent a nonredundant set of >45,000 human genes.

Precisely what constitutes a gene is somewhat contentious but a crude working definition might be: *a transcription unit plus associated regulatory sequences which together serve to specify both the sequence and the expression pattern of a protein product*. The term 'gene' cannot, however, be restricted to protein coding sequences since some genes (e.g. snRNA, rRNA, tRNA, XIST, H19, IPW) encode RNA molecules with a variety of biological functions and which are not translated into protein. A simple universally applicable definition of a gene is difficult to derive owing to the existence of exceptions to almost any rule that one might devise. Thus, some transcription units may encode multiple unrelated proteins with different functions as a result of alternative splicing. As a consequence, the notion of a gene becomes somewhat elastic. Some genes occur within the introns of other genes, for example OMG, EVI2A, EVI2B within the neurofibromatosis type 1 (NF1; 17q11.2) gene, F8A within the factor VIII (F8C; Xq28) gene, and U21 within the L5 genes of chickens and mammals. The genes of most known vertebrate small nucleolar mRNAs (snoRNAs) are located within the introns of other genes, two human examples being the RNE1 and RNE2 genes which are located within the mitotic regulator (CHC1; 1p36.1) and the 67 kDa laminin receptor (LAMR1; 3p21.3) genes, respectively. The realization that some genes reside within the introns of other genes makes the concept of the gene that much more diffuse.

Human genes can also overlap in a number of different ways. Thus, two genes encoding *erbA* homologues, *ear*-1 (THRAL) and *ear*-7 (THRA), located at the same locus on chromosome 17, possess overlapping exons but are transcribed from opposite DNA strands. Similarly, the tenascin-X (TNXA; 6p21.3) gene overlaps with the last exon of the cytochrome P450 (CYP21; 6p21.3) gene on the opposite DNA strand. Other examples of overlapping human genes transcribed from opposite DNA strands are provided by the PMS2 (7p22) gene and a gene encoding a 34.5 kDa polypeptide, the CD3 $\zeta/\eta/\theta$ (CD3Z) and Oct1 transcription factor (POU2F1) genes on lq22-q23, and the cytochrome *c* oxidase subunit X (COX10) gene and a partially characterized cDNA (C170RF1) on chromosome 17p12-p11.2. An example of overlapping human genes transcribed from the same strand is provided by the prematurely termed growth hormone gene-derived transcriptional activator (GHDTA) gene and the growth hormone 1 (GH1; 17q22-q24) gene. The GHTDA gene is transcribed from position -197 in the GH1 gene promoter and contains an open reading frame that extends from the ATG at –151 to a Stop codon in exon 2 of the GH1 gene. Another kind of overlap is exemplified by the transglutaminase 1 (TGM1; 14q11.2) gene some of whose *cis*-acting regulatory sequences reside within the transcribed portion of the functionally unrelated Rab geranylgeranyl transferase α subunit (RABGGTA; 14q11.2) gene located <2kb upstream of the TGM1 transcriptional initiation site. Clearly, the existence of overlapping genes can hamper any attempt to demarcate precisely and unambiguously where one gene ends the another begins.

Some genes such as the immunoglobulin and T-cell receptor genes differ in structure between different cell types. Which is the gene; that which is present in the germline or that which is rearranged to perform a specific function in the soma? Should distant regulatory sequences and protein binding sites that maintain chromosomal conformation be included within the boundaries of a gene? Indeed, if we are prepared to entertain radical redefinition of what constitutes a gene, should all parts of a gene necessarily be contiguous on the same DNA strand or even the same chromosome?

At the time of writing, the majority of human gene transcripts still remain to be characterized. However, Adams et al. (1995) attempted to ascribe functions to cDNAs by limited DNA sequence analysis and detection of homologies to known proteins. Of the human cDNAs studies by Adams et al. (1995), which probably represent no more than 10% of the total number, only eight of the corresponding genes were expressed in all 30 tissues examined. Some 227 genes were expressed in >20 tissues whilst some 4300 genes were found to be expressed in only one tissue. Clearly such data are extremely useful for any conceptual discussion of what we mean by tissue specificity of gene expression.

Table 1.1. Putative functions of a sample of human genes

Putative function	*Proportion of transcripts (%)*
Cell signalling	12
RNA synthesis/processing	6
Protein synthesis/processing	15
Metabolism	16
Cell division/DNA synthesis	4
Cell structure/mobility	8
Cell/organism defence/homeostasis	12
Unclassified	24

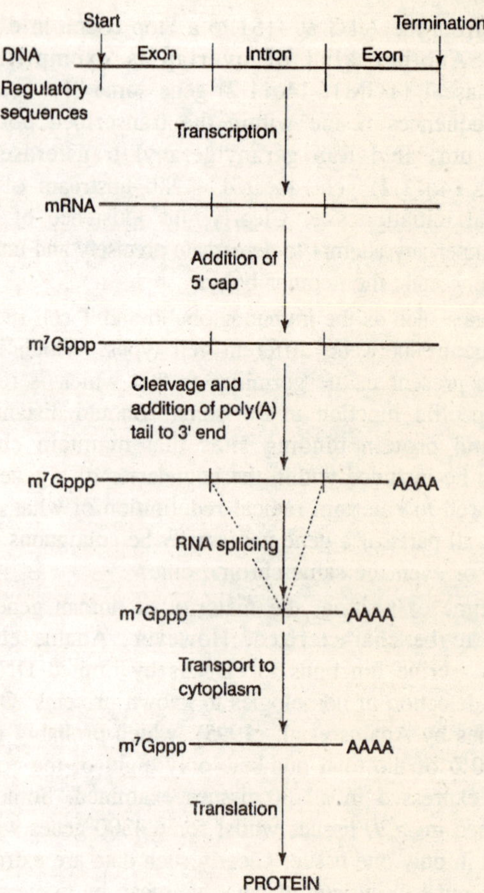

Fig. 1.2. Expression pathway of a human protein-coding gene.

Although the primary control of gene expression is usually exerted at the level of transcription, the regulation of gene expression may also occur at several other different stages in pathway including transcriptional activation, mRNA splicing, stability, export, translation (synthesis of the protein product), post-translational processing, and export of the mature protein. A variety of sequences within a typical gene region are require for normal and appropriate expression to occur. These are described briefly in subsequent sections.

Polymorphisms

The term *polymorphism* has been defined as a 'mendelian traits that exist in the population in at least two phenotypes, neither of which occurs at a frequency of less than 1%. Some DNA polymorphism are neutral single base-pair changes detected by virtue of the consequent introduction or removal of a restriction enzyme recognition site and are accordingly termed Restriction Fragment Length polymorphism (RFLPs). They are inherited as simple Mendelian traits since two alleles are generated as a consequence of the presence or absence of each restriction. Site. RFLps are not rare, being distributed throughout the genome at a frequency of between 1/200 and 1/1000 bp. Some 25% of single nucleotide polymorphism (SNPs) in higher primates occur in CpG dinucleotides consistent with a model of methylation-mediated deamination.

Not unexpectedly, the vast majority of polymorphisms occur in introns or intergenic regions rather than within coding sequences and may thus be expected to be neutral with respect to fitness. Those polymorphisms that occur either within coding regions or in the promoter region may however affect either the structure of function of the gene product or the expression of the gene and may therefore have the potential to be of phenotypic or even pathological significance. Those coding sequence polymorphism that alter the amino acid sequence of the encoded protein are found at a lower rate and with lower allele frequencies than silent substitutions. This probably reflects the action of negative selection on deleterious alleles during human evolution.

Some polymorphisms may be missense mutations, for example those underlying the Lewis Le alleles in the FUT3 gene or the Arg/Gln 353 polymorphism in the factor VII gene. Others are nonsense mutations that serve to inactivate the gene in question, for example the secretor se allele in the FUT2 gene (19cen-qter) present in 20% of the population. Further types of gene-associated polymorphism in the human genome include triplet repeat copy number, gene deletions, gene duplications, intragenic duplications, micro-insertions, inversions, gene fusion, and gene copy number. Various databases of human DNA polymorphism are available for online consultation: The Genome Database (http://gdbwww.gdb.org/), the Database of Single Nucleotide Polymorphisms (http://www.ncbi.nlm.nih.gov/SNP/), and the database of Human Genic Bi-Allelic Sequences (http://hgbase.interactiva.de/intro.html).

The mechanism by which polymorphisms are maintained in human populations are likely to be varied. The neutralist theory assumes no selection on the alleles of a polymorphic locus and the frequency of an allele may therefore increase simply by *genetic drift* (the change of allele frequency due to random sampling). Such 'transient polymorphism' often remain at a low frequency in the population before being lost or may instead increase in frequency under the influence of either genetic drift or positive selection until one allele reaches fixation. Most known polymorphisms are probably of this type. However, if the alternative alleles are not neutral with respect to fitness, the DNA polymorphism may be maintained by selection pressure, possibly overdominant selection ('balanced polymorphism'). Finally, a 'hitchhiker effect' may operate if the polymorphisms are closely linked to a locus which is itself under strong selection. In some special cases, such as polymorphisms with CpG dinucleotides, recurrent mutation may serve to maintain the allele frequency in the population.

The alleles of closely linked polymorphisms often occur in specific combinations or *haplotypes*. In the β-globin (HBB; 11p15) gene, for example, some haplotypes have been identified in Europeans, Asians, Blacks, and Chinese, indicating that their origin must have predated racial divergence. By contrast other haplotypes are population-specific or private'. In some cases, the number of different haplotypes can be quite high, the result of the shuffling of different single base-pair substitutions by such processes as recombination and gene conversion.

In their analysis of 8 kb of intronic sequence of the Duchenne muscular dystrophy (DMD; Xp21) genes of individuals from 13 different human populations identified 36 polymorphisms. Of these, 15 were shared among most of the populations screened, 13 were confined to Africans and four were confined to non-Africans. A detailed study of human DNA polymorphisms has also been performed on a 9734 bp region of the human lipoprotein lipase (LPL; 8p22) gene, comprising 8736 bp intronic sequence and 998 bp coding sequence, performed on 142 chromosomes from 71 individuals from three distinct populations: Finns, European-Americans and African-Americans. A total of 88 sites were found to vary (79 single base-pair substitutions and nine microdeletions or microinsertions) representing an average of one variable site every 500 bp. Only seven of the 88 variables sites were found in the coding region. At 34 of the 88 sites, the variation was found in only one of the three populations reflecting the differing population and mutational histories. A total of 88 unique haplotypes were identified in the 142 chromosomes sampled, probably a reflection of the complex historical interplay of recombination and mutation. Finally, a study of a total of 87 kb of human chromosome Xq22 revealed 102 polymorphisms, seven of which were shared by Europeans,

Ashkenazim and pygmies, two by pygmies and Europeans, and 19 by Ashkenazim and Europeans.

Some polymorphisms within human gene coding regions also occur in the orthologous genes of the great apes. In some cases, this may imply an origin for the polymorphism that predated the adaptive radiation of the higher primates ('trans-species polymorphism'). Perhaps the best example of this phenomenon is provided by the ABO blood group locus (ABO; 9q34.1-q34.2). In humans, there are four amino acid substitutions (in position 176, 235, 266 and 268) that distinguish the A- and B-transferase (respectively R, G, L, and G in A-transferase and G, S, M, and A in B-transferases) whilst the O allele is characterized by a single nucleotide (G_{261}) frameshift deletion that inactivates the protein. The gorilla and chimpanzee possess types B and A respectively whilst the baboon (*Papio cynocephalus*) exhibits both A and B. Certain alleles of the major histocompatibility complex (MHC) are also very ancient. Indeed, it would appear that some of the class I MHC allelic lineages are shared by human, chimpanzee and gorilla, implying that these polymorphisms were present in the common ancestor of the three species. Another possible example of trans-species polymorphism is to be found in the complement C4 genes (C4A, C4B; 6p21.3) of humans, chimpanzees and orangutans. Finally, trans-species polymorphism may also be present in the cone visual pigment genes, RCP and GCP (Xq28) but in this case, the evidence must be regarded as more equivocal. For polymorphisms to have survived the speciation process, they must have resisted fixation even during severe population bottlenecks. Neutral alleles would have been unlikely to survive the process of speciation and so the long -term persistence of polymorphisms must explicable by selection. In the case of MHC and ABO alleles, allelic lineages have predated human speciation and pathogen-driven *overdominant selection* has probably been the factor that has served to maintain the presence of alternative alleles. In other cases, it is conceivable that identical mutations could have occurred independently in the different lineages thereby mimicking true trans-species polymorphism.

Trans-specific polymorphism is likely to be the exception rather than the rule since, in the absence of positive selection, polymorphisms are most unlikely to survive speciation. Thus an average of $4N_e$ generations (N_e = effective population size) are required for a newly emerged allele to become fixed in a population as a consequence of random drift. Since until recently the effective population as a consequence of random drift. Since until recently the effective population size of modern humans has been ~10,000 individuals, most neutral human polymorphism may be expected to be less than 800,000 years old (assuming a 20 year generation time), rather less than the 5-6 Myrs that have elapsed since the divergence of the human and chimpanzee lineages. Hacia et al. (1999) studied the orthologous sequences from chimpanzees (both species) and gorilla for a total of 397 human single nucleotide polymorphisms (SNPs). They were able to determine the alleles in 214 of these SNP sites and of these, three segregated for the same nucleotides in both humans and pygmy chimpanzee whilst two segregated in humans and gorillas. However, 4/5 of the shared polymorphic sites occurred at hypermutable CpG dinucleotides suggesting that recurrent mutation rather than identity-by-descent was the cause of the shared polymorphism.

Functional Organization of Human Genes

Well before the organization of the human genome is known in its entirety, some trends governing the distribution of genes have become apparent and these are discussed briefly below.

Genes which encode the same product (e.g. ribosomal RNA (RNR), histones (H1F2, H1F3, H1F4, H2A, H2B, H3F2, H4F2), HLA, homeobox proteins (HOXA, HOXB, HOXC, and HOXD), immunoglobulins (IGK, IGL, IGH) are often clustered. However, these clusters are usually distributed between several different chromosomes (e.g. RNR, chromosomes 13, 14, 15, 21, and 22; histones, chromosomes 1, 6, and 12; HOX, chromosomes 2, 7, 12, and 17). Thus, multigene families have evolved by duplication and divergence but the duplicated copies may no longer be *syntenic* (i.e. linkage on the same chromosome conserved) owing to the translocation of discrete chromosomal regions during evolution. The likelihood of synteny may well be related to the time that has elapsed since duplication.

Genes which encode tissue-specific protein isoforms or isoenzymes are sometimes clustered (e.g. pancreatic (AMY2A, AMY2B) and salivary (AMY1A) amylase genes on chromosome 1p21) but sometimes not (e.g. cardiac (ACTC), skeletal muscle (ACTA1), smooth muscle, aorta (ACTA2) and smooth and muscle, enteric (ACTG2) a-actin genes which are located on chromosomes 15, 1, 10, and 2, respectively). Again, the time that has elapsed since the duplication event is likely to be an important factor in determining whether or not the duplicated genes have remained syntenic.

Genes encoding isozymes specific for different subcellular compartments are usually not syntenic [e.g. soluble/extracellular and mitochondrial forms of superoxide dismutase (SOD1/SOD3, SOD2 on chromosomes 21/4 and 6, respectively), aconitase (ACO1, ACO2 on chromosomes 9 and 22) and thymidine kinase (TK1, TK2 on chromosomes 17 and 16)].

Genes encoding enzymes catalyzing successive steps in a particular metabolic pathway are usually not syntenic. Thus the five enzymes of the urea cycle are encoded by genes (ARG1, ASL, ASS, CPS1, OTC) on chromosomes 6, 7, 9, 2 and X and four enzymes involved in galactose metabolism are encoded by genes (GALE, GALK1, GALK2, GALT) on chromosomes 1, 17, 15, and 9. However, there are exceptions to this rule: four genes encoding enzymes of the glycolytic pathway (TP11, GAPD, ENO2, LHHB) are located on the short arm of chromosome 12 in the region p13-p12. Similarly, the GDH and PGD genes encoding enzymes of the phosphogluconate pathway are encoded by linked genes on chromosome 1. The reasons for this syntenic organization, when it occurs, are usually unclear but we may nevertheless surmise the evolutionary history of the genes involved. Indeed, as early as 1945, Horowitz proposed that genes encoding enzymes of metabolic pathways could have arisen by serial duplication. His idea was that the protein that would eventually become the terminal enzyme of a given metabolic pathway would posses a binding site for the substrate that it used. A novel protein derived from the duplicated gene could still use this binding site for interaction with the same substrate molecule but could evolve the capability of producing it as a product from another source. Thus, the substrate for the terminal enzyme would become an intermediate in the developing metabolic pathway. In this way, the various enzymes of a pathway would evolve from each other in the reverse order to that in which they appear in the modern pathway. The synteny observed in the cases cited above might be a consequence of conservation resulting from a requirement for coordinate regulation of the loci concerned. Some of these principles also appear to hold for the genes encoding the enzymes of the coagulation cascade.

Genes encoding different subunits of a heteromeric protein are often not syntenic (e.g. the genes encoding α-globin (HBA1, HBA2, chromosome 16) and β-globin (HBB, chromosome 11), lactate dehydrogenases A (LDHA, chromosome 11) and B (LDHB, chromosome 12), factor XII subunits a (F13A, chromosome 6) and b (F13B, chromosome 1), the immunoglobulin light chains (IGK, IGL, chromosomes 2 and 22), and heavy chains (IGH, chromosome 14). However, several cases of synteny are known e.g. the genes encoding the three chains of fibrinogen (FGA, FGB, FGG, all closely linked on chromosome 4), the α- and β-chains of C4b-binding protein (C4BPA and C4BPB, closely linked on chromosome lq32), the complement component 1Q α- and β-chains (C1QA, C1QB, linked on chromosome 1p) and the platelet membrane glycoproteins IIb and IIIa (ITGA2B, ITGA3, closely linked on chromosome 117q21-q22). The clustering of these genes may be important for their coordinate regulation by common control element. The various subunits of the T-cell antigen receptor are intriguing: the α- and δ-subunits are encoding by genes (TCRA, TCRD) on chromosome 14, the β- and γ-subunits genes (TCRB, TCRG) are on chromosome 7 whereas the ε-subunit genes (TCRE) lies on chromosome 11. Whilst synteny probably implies a common evolutionary origin, lack of synteny does not necessarily argue against it.

Clustering of genes of similar function and common evolutionary origin is common, for example the genes encoding blood coagulation factors VII (F7) and X (F10) on chromosome 13q34, the γ-crystallin (CRYG) gene cluster on chromosome 2q33 and the six alcohol dehydrogenase (ADH1, ADH2, ADH3, ADH4, ADH5, ADH7) genes on chromosome 4q22.

Genes do not usually exhibit chromosomal clustering with respect to the structure/function of particular organs or subcellular organelles (e.g. mitochondria). However, various genes encoding proteins expressed in the course of epidermal differentiation [involucrin (IVL), loricrin, (LOR), filaggrin (FLG), the small proline-rich proteins (SPRR1A, SPRR1B, SPRR1B, SPRR2A and SPRR3), trichohyalin (THH) are clustered together on chromosome lq21 thereby betraying their common evolutionary origin. A considerable number of the genes encoding various cytokines (including several hematopoietic growth factors) and their receptors are clustered on the long arm of chromosome 5: granulocyte-macrophage colony-stimulating factor (GMCSF), macrophage colony-stimulating factor (CSF2), the CSF1 receptor (CSF1R, colony-stimulating factor-1 receptor, also known as *c-fms*), interleukins 3, 4, 5, 9, 12B, and 13 (IL3, IL4, IL5, IL9, IL12B, IL13), platelet-derived growth factor receptor-β (PDGFRB), acidic fibroblast growth factor (FGF1) and fibroblast growth receptor 4 (FGFR4). The genes encoding the IL3 receptor α-chain (IL3RA) and the GMCSF receptor α-chain (CSF2RA) both map to the pseudoautosomal region of the sex chromosomes. Synteny betrays the common evolutionary origin of the genes as well as the probable mechanism—tandem duplication.

Genes encoding ligands and their associated receptors are sometimes syntenic, for example the genes encoding transferrin (TF) and its receptor (TFRC) arc both located on chromosome 3q whilst the genes encoding apolipoprotein E (APOE) and the low density lipoprotein receptor (LDLR) are both located on chromosome 19. However, not surprisingly, this is far from always the case, for example insulin (INS, chromosome 11) and insulin receptor (INSR, chrommosome 19); epidermal growth factor (EGF, chromosome 4), epidermal growth factor receptor (EGFR, chromosome 7); growth hormone (GH1, GH2, chromosome 17) and growth hormone receptor (GHR, chromosome 5); interferon α, β, γ and ω1 (IFNA, IFNB1, IFNG, IFNW1, chromosomes 9, 8, and 9, 12, 9), interferon receptors $\alpha/\beta/\omega$ and γ (IFNAR1, IFNGR1, chromosome 6).

The linear order of members of a family of related genes can reflect the order in which they become activated during development, for example the HBE1 (embryonic), HBG2, HBG1 (fetal), HBD, HBB (postnatal) genes of the human β-globin cluster. The expression of these genes is controlled by an upstream locus control region (LCR) and correct gene order is required for the normal temporal pattern of developmental expression. Another example of this phenomenon is provided by the HOX genes which are organized chromosomally according to their order of expression. Intriguingly, the encoded Hox proteins vary in their affinity for their target DNA sequences and these affinities also correlate with the linear order of the genes. By contrast, the order of the human ADH genes on chromosome 4q21, 5'-ADH3-ADH2-ADH1-3', is opposite to their order of transcriptional activation in hepatic development although the significance of this observation, if any, is unknown. The reflection of temporal order of expression in the physical order on the chromosome is not however a universal phenomenon as is evidenced by the human and murine myosin heavy chain (MYH) gene clusters on chromosomes 17 and 11, respectively. Finally, it may be significant that the order of four human mucin genes (MUC2, MUC5AC, MUC5B, MUC6) on chromosome 11p15 corresponds to their order in terms of the anterior-posterior axis of the epithelial areas where they are preferentially expressed.

Pseudogenes

Pseudogenes are DNA sequences which are closely related to functional genes but which are incapable of encoding a protein product on account of the presence of deletions, insertions and nonsense mutations which abolish the reading frame or otherwise prevent gene expression. Some human pseudogenes are transcribed but these transcripts are not translated. There are two major types of pseudogene:

the first arises through the duplication and subsequent inactivation of a gene. This type of pseudogene retains the exon/intron organization of the parental gene and are often closely linked to the parental gene. Examples include the pseudogenes in the α- and β-globin clusters. The second type of pseudogene contains only the exons of the parental gene, usually possess a poly(A) tail at the 3' end and are dispersed randomly in the genome. These *processed* genes are thought to have originated as mRNAs which have then become integrated into the genome by *retrotransposition* (i.e. the reverse transcription of the mRNA and the integration of the resulting cDNA).

Pseudogenes are relatively common in the human genome and may be especially prevalent in multigene families (e.g. β-globin, actin, HLA, interferons, snRNAs, keratins, T-cell receptors, immunoglobulin gene clusters). However, single copy genes may also have multiple pseudogenes (e.g. prohibitin, PHB, four pseudogenes; argininosuccinate synthetase, ASS, 14 pseudogenes).

Promoter Elements

The archetypal gene contains promoter elements upstream (5') of the transcriptional initiation site (i.e. the beginning of the mRNA) which serve to specify the temporal and spatial pattern of expression of the downstream gene and define its potential for induction by external stimuli. Some genes contain multiple alternative promoters which are utilized in a tissue-specific fashion (e.g. DMD). Some promoters may be located within an intron (e.g. within the first intron of the human ADA genes). Some promoters may even be found buried within the introns of other genes (e.g. an element of the human CYP21 gene promoter lies within intron 35 of the C4A gene).

HSPA1A

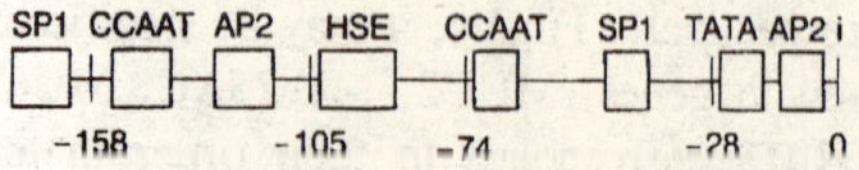

MT2A

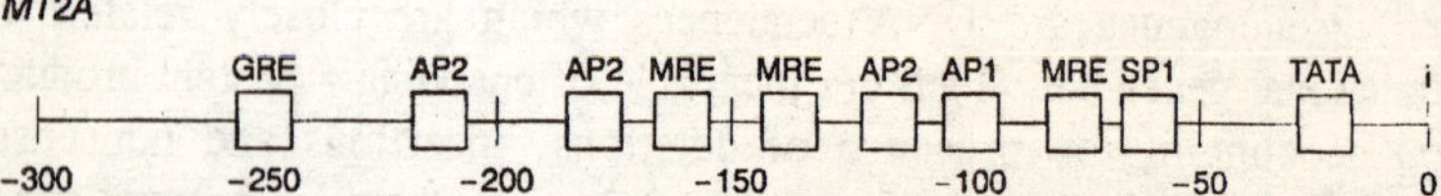

Fig. 1.3. Transcriptional control elements upstream of the transcriptional initiation site in the human heat shock protein 1 (HSPA1A; 6p21.3) and metallothionein 2A (MT2A; 16q13) genes.

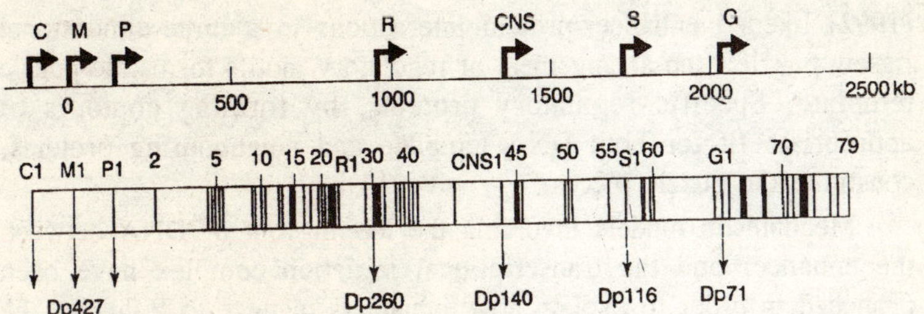

Fig. 1.4. Alternative promoter usage in the human dystrophin (DMD) gene.

The transcriptional initiation site is usually preceded by *constitutive promoter elements* of defined sequence, for example the TATA box (TATAAA; 25-30 bp 5' to the cap site), initiator element (Py Py A_{+1} N T/A Py Py) which serves as a functional analogue of the TATA box in TATA-less promoters and the CCAAT motif (~90 bp 5' to the cap site) which potentiate a basal level of gene expression. Further upstream regulatory motifs bind tissue-specific transcription factors, some binding different factors in different tissues. *Response elements*, as the name suggests, are able to confer transcriptional responsiveness to various external trigger stimuli such as temperature hormones and growth factors. In this way, gene expression is rendered responsive to both the internal (cellular) and external (to the organism) environment. Together, these upstream DNA sequences are involved in controlling gene regulation and induction and conferring tissue-specificity of expression.

A large number of DNA sequence motifs have now been identified within gene promoters which represent binding sits for DNA-binding proteins. These protein-DNA interactions are required to confer appropriate regulation upon the genes bearing them. Removal of these DNA sequence motifs abolishes the gene's specific pattern of spatial and/or temporal expression.

Enhancers

Enhancers are DNA sequences that are present 5' or 3' to a gene (or within an exon or an intron) and which are capable of activating the transcription of the gene in a tissue-specific fashion, independently of their orientation and distance from the correct initiation site. Collins et al. (1998) reported the unusual case of the mouse thrombospondin 3 (Thbs3) enhancer that is located far upstream of the Thbs3 gene, within intron 6 of the divergently transcribed metaxin (Mtx) gene.

Enhancers function by acting as templates for the assembly of multiprotein complexes on the gene promoter. Thompson and McKnight

(1992) likened enhancer-protein interactions to a three-dimensional jigsaw puzzle: 'the arrangement of regulatory motifs forms the puzzle template. Specific regulatory proteins, by forming contours of appropriate fit for both DNA template and neighbouring proteins, constitute the puzzle pieces.'

Mechanistic models involving the looping out of DNA between the enhancer and the transcriptional initiation complex have been proposed in order to explain how enhancers manage to influence the activity of their target promoters at considerable distances.

Negative Regulatory Elements

The negative regulation of transcriptional by *repressors* or *silencers* is now so well documented, that it is likely that many if not most genes are subject to their inhibitory influence. Indeed, every gene promoter region is likely to possess its own unique combination of positive and negative regulatory elements which serves to determine its temporal and spatial pattern of expression. These elements permit the binding of a specific set of DNA-binding proteins which are thereby brought into sufficiently close proximity so as to allow their interaction both with each other and with the RNA polymerase in order to influence transcription either positively or negatively. Negative regulation therefore serves to prevent the expression of a gene in an inappropriate tissue or at an inappropriate time or at an inappropriate level. They also potentiate the down-regulation of the expression of a gene following its transient induction. Negative regulatory elements have been found not merely in the promoter regions of human genes as in the α1-antitrypsin (PI; 14q32.1) gene but also in the first exon [osteocalcin (BGLAP; 1q25-q31) gene], the first intron [(CD4; 12pter-p12) gene], or even the third intron [type IV collagen (COL4A2; 13q34) gene] of a gene.

Locus Control Regions

Regulatory elements that exert their effects on the expression of downstream genes over great distances have been described in the human α-globin (HBA1; 16p13.3), β-globin (HBB; 11p15), growth hormone (GH1; 17q22-q24) and red/green cone pigment (RCP, GCP; Xq28) genes among others. These are known as *locus control regions* (LCRs). The LCR located 40 kb upstream of the HBB gene is essential for the high level, tissue-specific expression of the HBB gene. It is believed to contain an enhancer capable of controlling the replication timing of the β-globin gene cluster, organizing it into an active chromatin domain and directing the expression of downstream sequences in an erythroid-specific and developmental stage-specific fashion. Although conserved between mammals, little homology is apparent between mammalian and avian β-globin LCR regions.

Trans-acting Protein Factors

As we have seen, the transcriptional activation of eukaryotic genes is made possible by the interaction of *trans*-acting protein factors with *cis*-acting DNA sequence motifs including enhancers. These transcription factors typically contain a sequence-specific DNA-binding domain, a multimerization domain which allows the formation of either homomultimers or heteromultimers, and a transcriptional activation domain. These domains can be combined in modular fashion to generate an array of different transcription factors.

It is the *cis*-acting DNA sequences within a gene promoter that allow transcription factors to be brought into close proximity so that they may either interact with each other in the transcriptional initiation complex or combine together in an enhancer complex. No one enhancer-binding protein can act on its own, rather it must act in concert with other enhancer-binding proteins. For example, one factor may induce a bend in the DNA thereby promoting the interaction of two already bound proteins with each other. Once the enhancer complex is assembled it must be able to interact either directly or indirectly with the basal transcription apparatus via its activation domain.

Transcription factors can be grouped into families of related proteins whose relatedness extends to homology in their DNA-binding domains and therefore an ability to bind to related DNA sequences. Specific transcription factors can thus bind to more than one DNA sequence. Conversely, a single DNA sequence motif may sometimes be bound by more than one transcription factor. DNA-binding domains fall into one of four main groups defined by homologous amino acid sequences that give rise to a particular structure capable of binding DNA; homeodomain, zinc finger, leucine zipper and helix-loop-helix. These domains usually bind the negatively charged DNA molecule through basic amino acid residues. Activation domains of transcription factors also come in four types: rich in either glutamine, proline, serine/threonine, or acidic amino acids. Examples of these types are SP1, CTF/NF-1, Pit-1, and GATA-1, respectively.

Transcriptional repressor proteins may bind directly to DNA or exert their repressive effects indirectly by interacting with basal transcription factors, transcriptional activators and co-repressor proteins to inhibit transcription by RNA polymerases.

The study DNA-protein interactions has been enormously facilitated by the use of two techniques: gel retardation analysis (also termed band or mobility shift assays; Dent and Latchman, 1993) and DNase I footprinting. The former technique is extremely useful for searching for DNA-binding proteins in crude nuclear extracts whereas the latter method provides information as to the precise location of the binding site on the DNA sequence under study.

5' and 3' Untranslated Regions

The presence of the 5' untranslated region (5' UTR; the sequence lying between the transcriptional initiation site and the translational start codon, ATG) is often essential for the normal expression of a gene. Sequences in the 5' UTRs of various genes are thought to play a role in controlling the translation of the encoded mRNA. Perhaps the best characterized post transcriptional control mechanism involving the 5' UTR is that of the iron response element (IRE). The IRE is found in the 5' UTRs of several human genes [e.g. ferritin (FTH1), transferrin (TF), erythroid 5-aminolevulinate acid synthase (ALAS1)] and is capable of adopting a stem-loop structure that interacts with a cytosolic RNA-binding protein thereby inhibiting mRNA translation. The post-transcriptional regulation of several other human genes [e.g. transforming growth factor-β1 (TGFB1) and basic fibroblast growth factor (FGF2) also appears to involve regulatory elements that modulate the efficiency of translation.

Zhang (1998) examined the nature of translational start codons in human genes. The relative frequencies of start codons occurring as the first, second, third or fourth ATG codons in the reading frame were 474, 51, 5, and 0, respectively. There are, however, quite a few examples of more than one ATG being used in the same gene. Thus, the alternative use of two ATG codons 84 bp apart in the human peroxisome proliferator-activated receptor (PPARG; 3p25) gene serves to generate two distinct protein isoforms that differ in length by 28 amino acids at the amino terminal end of the protein. The alternative use of different ATG initiation codons has also been reported for the human von Hippel Lindau (VHL; 3p25-p26) gene which leads to the production of two distinct protein products that differ in length by 53 amino acids.

Zhang (1998) reported that the ratio of the frequencies of stop codons TAA, TAG and TGA in human genes was about 1:1:2. The 3' untranslated region is that region of a gene which lies downstream of the stop codon. The AATAAA sequence located downstream of the stop codon (TAA) of ~90% of genes (i.e. in the 3' untranslated region) controls cleavage/polyadenylation (addition of a poly A tail 10-30 bp downstream) which is vital for mRNA stability. The AAUAAA motif is thought to interact with at least two RNA-binding proteins, cleavage-polyadenylation specificity factor (CPSF) and cleavage-stimulatory factors (CstF), which are thought to promote assembly of the polyadenylation complex. Some genes possess multiple *polyadenylation sites* which can be used alternatively.

Other sequences have been implicated in determining mRNA stability, for example a GU-rich motif (consensus sequence, YGUGUUYY), present 20-40 bases downstream of the cleavage site in ~70% of mammalian pre-mRNAs, that plays a crucial role in 3' end formation. A further U-rich element, upstream of the AAUAAA polyadenylation motif is also involved. Some short-lived mRNAs possess AU-rich regions which specify instability. Still other sequences may be involved in mRNA export, nucleocytoplasmic transfer, intracellular localization and in promoting translational efficiency. A wide variety of RNA-binding proteins are now known which play an important role in the post-transcriptional regulation of gene expression.

Boundary Elements

The position and orientation independence of enhancers and LCRs clearly raises the potential problem of the inappropriate activation of promoters from neighbouring genes. Sequences which constrain the activity of enhancers were originally reported in mice and *Drosophila*. Termed 'boundary elements' or 'insulators', these sequences serve to insulate a gene from the effects of either enhancer or suppressor elements emanating from the surrounding chromatin. This insulator function appears to work either by blocking interactions with other sequences past the boundary element and/or by limiting the influence of an enhancer to the locality of its target gene(s). Boundary elements therefore serve as functional barriers when inserted between an enhancer and its downstream reporter gene. However, bracketing an enhancer–gene combination with boundary elements serves to confine the enhancer activity to that gene's promoter such that its expression is maintained or even increased. MARs (*matrix attachment regions*) may act as boundary elements as evidenced by their ability to establish transcriptional domains around transgenes and act as buffers to shield these transgenes from position effects.

A boundary element has been described 5' to the LCR of the chicken β-globin gene cluster (hypersensitive site 4); it serves to insulate genes 5' to it from the influence of the LCR, a function which it also manifests in both human and *Drosophila* cells. DNA around the DNase-I hypersensitive site 5 of the human β-globin gene appears to function in a similar fashion.

Another possible example of a boundary element in humans has been noted upstream of the coagulation factor X (F10; 13q34) gene. This gene lies 2.8 kb downstream of the gene (F7) encoding the homologous clotting protein, factor VII. Three positive regulatory elements (FXP3, FXP2, and FXP1) have been characterized upstream of the F10 gene. FXP1 and FXP2 act in an orientation- and position-independent fashion whilst FXP1 and FXP3 are responsible for directing liver-specific gene expression. The putative boundary element just upstream of the FXP3 sequence is thought to prevent transcriptional activation of the F7 gene by the F10 gene enhancers.

Sequence Motifs Involved in mRNA Splicing and Processing

One of the characteristics of eukaryotic genes that distinguishes them from their prokaryotic counterparts is the production of large pre mRNAs which contain intervening non-coding sequences (introns) that are removed by a highly accurate cleavage/ligation reaction known as splicing before the mRNA is transported to the cytoplasm for translation. Splicing not only permits the removal of introns from the primary transcript but also allows the generation of different mRNAs from the same gene by *alternative splicing*, an important mechanism for tissue-specific or developmental regulation of gene expression and a very economical means of generating biological diversity. Alternative splicing may be regulated by variation in the intracellular levels of antagonistic splicing factors.

The splicing of a eukaryotic mRNA appears to occur as a two-stage process. In the case of a simple two exon gene, the pre-mRNA is first cleaved at the 5' (donor) splice site to generate two splicing intermediates, an exon-containing RNA species and a lariat RNA species containing the second exon plus intervening intron. Cleavage at the 3' (acceptor) splice site and ligation of the exons then occurs resulting in the excision of the intervening intron in the form of a lariat. Splicing efficiency is critically dependent upon the accuracy of cleavage and rejoining. This accuracy appears to be determined, at least in part, by the virtually invariant GT and AG dinucleotides present at the 5' and 3' exon/intron junctions respectively. However, more extensive consensus sequences spanning the 5' and 3' splice junctions are evident and the coding sequence flanking intron junctions exhibits

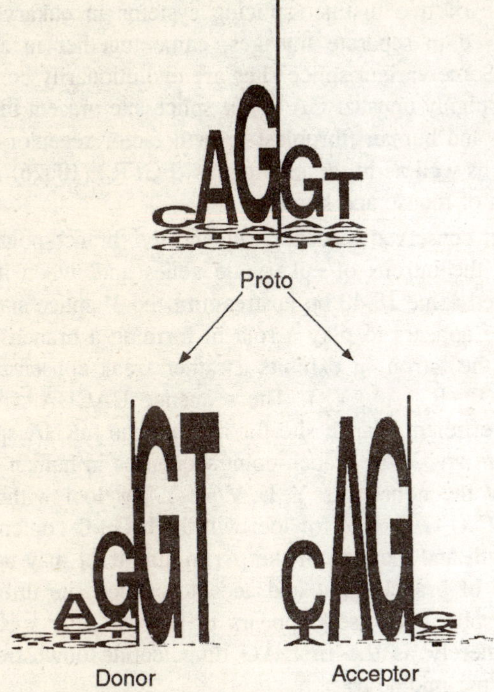

Fig. 1.5. Putative schema for the evolution of donor and acceptor splice sites from a single 'proto-splice site' ancestor.

some degree of conservation. More recently, Zhange (1998) has proposed AG | GTRAGT as a consensus sequence for donor splice sites and $(Y)_n$NCAG |G (where *n* has a mean of nine) as a consensus for acceptor splice sites. Stephens and Schneider (1992) noted the similarity between the donor and acceptor sites. Stephens and Schneider (1992) noted the similarity between the donor and acceptor sites and suggested that these junctions may have been derived from a common 'proto-splice site' ancestor. During evolution, the emphasis of the sequence information at each site has shifted to the intronic side of the junction.

A few human nuclear genes possess introns with noncanonical terminal dinucleotide sequence (e.g. 'AT-AC introns'); these include the transcription factor E2F1 (20q11.2) gene, the cartilage matrix protein (MATN1; 1p25) gene, the Hermansky-Pudlak syndrome (HPS; 10q23) gene and the paralogous sodium channel α-subunit genes (SCN4A, 17q23.1-q25.3; SCN5A, 3p24-p21; SCN8A, 12q13). The removal of AT-AC introns requires two low-abundance snRNAs, U4atac and U6atac, not found in the major spliceosome. There is now good evidence for the existence of two distinct splicing systems in eukaryotes which, having evolved in separate lineages, came together in a eukaryotic progenitor. Some variant splice sites are evolutionarily conserved, for example the highly unusual GA donor splice site present in exon 10 of the *Xenopus* and human fibroblast growth factor receptor 1 (FGFR1; 8p11) genes as well as in the paralogous (FGFR2 (10q26) and FGFR3 (4p16) genes of mouse and human.

A further conserved sequence element, the 'branch-point', has been identified in the introns of eukaryotic genes and, when it occurs, is usually located some 18-40 bp upstream of the 3' splice site. Although this sequence appears to play a role in forming a branch with the 5' terminus of the intron, it exhibits a rather weak consensus sequence (Y_{81} N Y_{100} T_{87} R_{81} A_{100} Y_{94}). The sequence UACUAAC appears to be the most efficient branch site for mammalian mRNA splicing both *in vitro* and *in vivo*. The branch-point sequences in human can best be described by the consensus Y U V A Y for loci with low G+C content and C U G/C A Y for loci with high G+C content. Whereas both the length and location of the pyrimidine tract may be important determinants of branch-point and acceptor splice site utilization, the 3' acceptor splice site itself appears to possess little specificity and may serve merely as the first AG dinucleotide downstream of the branch-point/pyrimidine tract.

A further category of sequences required for alternative splicing are the *splicing enhancers*. Several types of splicing enhancer have been recognized. These sequences occur in either exons or introns and serve to promote the use of neighbouring weak 5' or 3' splice sites. They do this through the binding of serine arginine (SR) proteins, a family of modular splicing factors which contain one or more RNA-binding domains. Splicing occurs within the *spliceosome*, a complex assembly of small ribonucleoprotein particles (snRNPs) composed of a variety of snRNAs and associated proteins. The pre-mRNA is folded in such a way that splice sites are optimally aligned for cleavage and ligation. In this process, the snRNA play a vital role. Our current understanding of mRNA splicing suggests that the formation of the 5' splice site complex is contingent upon the prior formation of the 3' splice site complex. Spliceosomal structure and function has been conserved in eukaryotes from yeast to humans.

Polycistronic and Polyprotein Genes

In most cases, human genes adhere to the principle of 'one gene, one polypeptide'. However, some multi-chain protein genes encode more than one polypeptide. In such cases, of a polypeptide precursor followed by association of the constituent chains. Thus, the human insulin (INS; 11p15) gene encodes an A chain, a B chain, and a connecting peptide which is thought to be important in maintaining the conformation of the two chains. More often than not, however, multichain proteins are encoded by different genes, for example the α- and β-globins. Nevertheless, even although the microtubule-associated proteins 1A and 1B (MAP1A, 15q13-qter; MAP1B, 5q13) are encoded by different genes, both genes encode heavy and light chains that must be proteolytically processed from a precursor polypeptide.

Several human genes are *polycistronic* in that they encode different peptides with different functions, for example calcitonin/calcitonin gene-related peptide α (CALCA, 11p15), pancreatic polypeptide/pancreatic icosapeptide (PPY; 17p11-qter) and vasoactive intestinal polypeptide/PHM27 (VIP; 6q26-q27). A further putative polycistronic gene has been characterized in both mouse and human: the growth/differentiation factor-1 (GDF1) gene is cotranscribed with a gene of unknown function (Uog1) separated from each other by a 269 bp intercistronic region. Finally, Reiss et al. (1988) described the polycistronic human molybdopterin synthase (MOCOD; 6p21.3) gene which contains two overlapping open reading frames (ORF) encoding distinct polypeptides required for the synthesis of molybdenum cofactor. The two polypeptides are encoded by the same 3.2 kb mRNA, the first by exons 1-9, the second by exon 10. Stallmeyer et al. (1999) have demonstrated that each ORF is translated independently.

More dramatic is the case of the human glycinamide ribonucleotide synthetase (GART; 21q22.1) gene which generates an mRNA encoding a trifunctional protein with three distinct enzymatic activities (glycinamide ribonucleotide synthetase, glycinamide ribonucleotide formyltransferase and aminoimidazole ribonucleotide synthetase which encode respectively the second, third and fifth enzymatic steps in the *de novo* synthesis of purines). These activities are separate in bacteria and partly separate in yeast implying that ancient recombination events may have fused and coding regions together. Fusion could have occurred as a means to allow coordinate regulation of the three enzymatic activities.

One way in which polycistronic mRNAs might have arisen during evolution is by inefficient RNA polymerase termination of transcription of one mRNA species allowing the cotranscription of a second mRNA species from a downstream gene. Such a phenomenon has been shown to occur in normal human cells and involves the *fusion splicing* of two adjacent (9p13) genes, those encoding galactose-1-phosphate uridylyltransferase (GALT) and interleukin-11 receptor α-chain (IL11RA).

Some human genes encode *polyproteins* and comprise multiple tandem repeats of the same coding region on the same transcript, for example those encoding the ubiquitins (UBA52, 19p13; UBB, 17p11-p12; UBC, 12q24) and filaggrin (FLG; 1q21).

DNA Methylation

5-methylcytosine (5mC) is the most common form of DNA modification in eukaryotic genomes. Soon after DNA synthesis is complete, target cytosines are modified by a DNA methyltransferase using S-adenosylmethionine as methyl donor. In humans, between 70% and 90% of 5 mC occurs in CpG dinucleotides, the majority of which appear to be methylated.

Whereas the vertebrate genome is heavily methylated, methylation is virtually undetectable in insects and other arthropods. An intermediate level of methylation is exhibited by the echinoderms, coelenterates and molluscs whose genomes are characterized by the presence of long methylated and unmethylated tracts. The transition from a fractional to global methylation pattern appears to have occurred close to the origin of the vertebrates since the cephalochordate *Amphioxus* exhibits a typically invertebrate pattern of genome methylation whereas the jawless vertebrates hagfish and lamprey possess a vertebrate pattern. The increase in size of the methylated compartment characteristic of vertebrate genomes correlates with the sharp increase in gene number during the invertebrate-vertebrate transition. DNA methylation may thus have been recruited as a transcriptional regulatory mechanism.

DNA methylation is essential for normal mammalian development. It is thought to play a role in both gene regulation and imprinting may serve as cue for strand specificity in DNA replication and repair and could conceivably serve as a self-defence mechanism to silence transposable elements and proviral DNAs integrated into the genome during evolution. This post-synthetic modification occurs almost exclusively in CpG dinucleotides of which between 60% and 90% are methylated in mammalian tissues.

Although it is as yet unclear how tissue-specific methylation patterns are established, they are nevertheless heritable and reproducible after transmission through the germline. The establishment of cell-type specific methylation patterns in both the soma and the germline begins with global methylation of non CpG island sequences in the embryo whilst the final methylation patterns are determined by a specific and highly regulated process of demethylation.

In humans, differences in site-specific methylation patterns between different ethnic groups have not so far been apparent in studies of a variety of different types of DNA sequence. In keeping with these reports, Millar et al. (1998) found no evidence in sperm DNA for variable methylation status in the factor VIII (F8C; Xq28) gene between European Caucasians and Asians. However, the methylation status of specific CpG sites in the F8C gene did exhibit significant inter-individual variation.

CpG Islands

Spatially, the distribution of CpG appears to be nonrandom in the human genome; about 1% of the genome consists of stretches very rich in CpG which together account for roughly 15% of all CpG dinucleotides. In contrast to most of the scattered CpG dinucleotides, these *CpG islands* represent unmethylated domains and comprise ~50% of all unmethylated CpGs in the genome. CpG islands occur, on average, every 100 kilobases in the murine genome and, in human, are often located immediately 5' to gene coding regions. The methylation of CpG islands may modulate gene expression through its influence on chromatin organization.

Not all vertebrate genes, however, possess CpG islands and many are partially or even heavily methylated. In general, gene promoters containing CpG islands are unmethylated regardless of expression

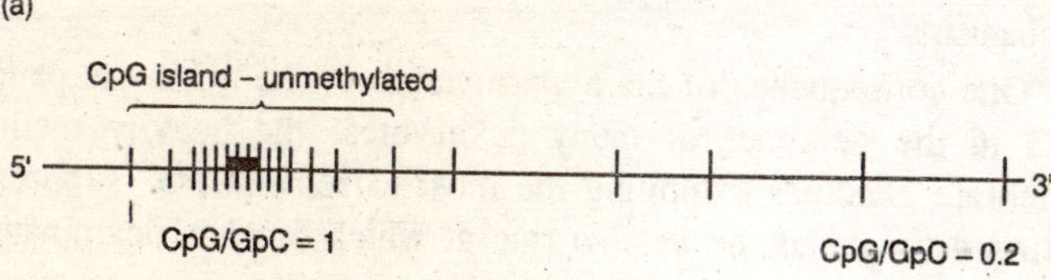

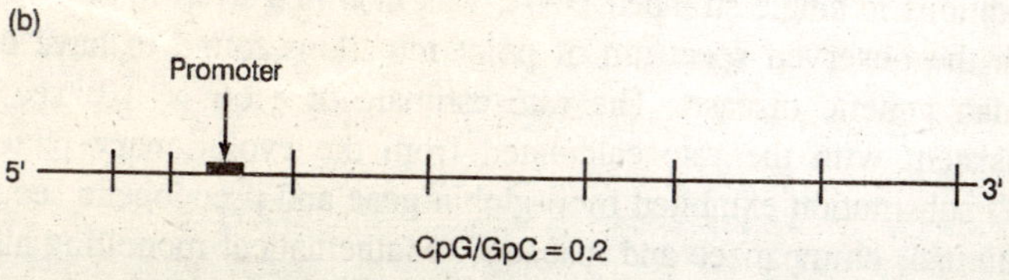

Fig. 1.6. The distribution of CpG within genes. (a) A typical CpG island containing gene as found in many 'housekeeping' genes. (b) A typical tissue-specific gene lacking a CpG island.

whereas promoters lacking CpG islands tend to lose their methylation upon transcription. CpG islands sometimes also occur within the coding regions of genes as in the case of the human apolipoprotein E (APOE; 19q13) gene in which exon 4 constitutes a 940 bp CpG island. Other examples of human genes with CpG islands within their coding regions are the CDKN2A (9p21), MYOD1 (11p15.4), and PAX6 (11p13) genes.

CpG islands are predominantly found in early replicating (R band) as opposed to late replicating (G band) regions of the genome. They may be considered to be the last remnants of the unmethylated domains that once dominated the vertebrate genome. The evolution of the heavily methylated vertebrate genome has been accompanied by a progressive loss of CpG dinucleotides as a direct consequence of their methylation in the germline. Although CpG islands are usually unmethylated and therefore relatively immune to mutational decay, there is nevertheless some evidence for their gradual erosion over evolutionary time.

CpG Suppression in the Vertebrate Genome and its Origin

Methylation of cytosine results in a high level of mutation due to the propensity of 5mC to undergo deamination to form thymidine. Deamination of 5mC probably does not occur during the enzymatic replication of the methylation pattern which appears to be a high fidelity process. Indeed, 5mC deamination probably occurs with the same frequency as the deamination of cytosine to uracil. However, whereas uracil DNA glycosylase activity in eukaryotic cells is able to recognize and excise uracil, thymine being a 'normal' DNA base is though to be both less readily detectable and removable by cellular DNA repair mechanisms.

One consequence of the hypermutability of 5mC is the paucity of CpG in the genomes of many eukaryotes, the heavily methylated vertebrate genomes exhibiting the most extreme 'CpG suppression'. A first estimate of the *in vivo* rate at which 5mC is deaminated and fixed as thymidine was arrived at by extrapolation from *in vitro* data. To this end, the deamination rate of 5mC as measured under laboratory conditions in single stranded DNA, was modified so as to be consistent with the observed spectrum of point mutations found to have caused human genetic disease. The rate estimate of $1.66 \times 10^{-16} sec^{-1}$ was consistent with the rate calculated from the evolutionary pattern of CpG substitution exhibited by β-globin gene and pseudogene sequences in human, chimpanzee and macaque. Mathematical modelling allowed a detailed study of the dynamics underlying the CpG suppression currently found in the bulk DNA of vertebrate genomes. It was inferred that the time span required in order to create the currently observed CpG frequency (0.01) was 50 to 100 Myrs. On the other hand, the process of CpG loss must have lasted for approximately 450 Myrs in order for the mononucleotide frequencies to have attained their present levels. This time span corresponds closely to the estimated time since the emergence and adaptive radiation of the vertebrates and thus coincides with the probable advent of heavily methylated genomes. These data are therefore consistent with patterns of vertebrate gene methylation having been comparatively stable over relatively long periods of evolutionary time.

CpG Dinucleotide as a Mutation Hotspot

CpG has been found to be a hotspot for mutation in a wide range of different human genes. Of the single base-pair substitutions so far reported to cause human genetic disease ~30% involve a CpG dinucleotide, and ~79% (~23% of the total) of these are either C→T or G→A transitions. These data imply that the rate at which CpG mutates to either TpG or CpA is some five-fold higher than the basal mutation rate per nucleotide. Methylation-mediated deamination of 5mC therefore represents a major cause of gene mutability leading to human genetic disease.

The proportion of human gene mutations compatible with a model of methylation-mediated deamination (CG→TG, CA) is however very much an average figure and provides no information on individual genes. When these values are recalculated on a gene-specific basis, considerable variation becomes apparent. The frequency of CG→TG and CG→CA mutations may be smaller than 10% [HBB (4%), HPRT1 (5%), TTR (9%)] or larger than 50% [MYH7 (67%), ADA (63%), RB1 (57%)]. In the assumed absence of a detection bias, we may surmise that this variation is due either to differences in (a) germline DNA methylation and/or (b) relative intragenic CpG frequency [itself dependent upon (a)]. Imbalances in the propensity to transition at CpG dinucleotides exist even between the two DNA strands. For the human hypoxanthine phosphoribosyltransferase (HPRT1) gene, Skandais et al. (1994) have noted a significant strand bias in mutations recovered at CpG sites, thereby confirming our own findings.

CpG hypermutability in inherited disease implies that the CpG sites in question are methylated in the germline thereby rendering them prone to 5mC deamination. However, on its own, CpG hyperpmutability still represents very indirect evidence for CpG methylation. That 5mC deamination is itself directly responsible for these mutational events is evidenced by the fact that several cytosine residues known to have undergone a germline mutation in the low density lipoprotein receptor (LDLR; hypercholesterolemia), p53 (TP53; various types of tumor), factor VIII (F8C; hemophilia A), and neurofibromatosis type 1 (NF1) genes are indeed methylated in sperm. As yet, however, these are the only studies to have attempted to correlate CpG hypermutability with DNA methylation directly for specific CpG dinucleotides.

Imprinting and Impr inted Genes

DNA methylation is thought to be involved in *imprinting* which was defined by Monk (1988) as the 'differential modification of the maternal and paternal contributions to the zygote, resulting in the differential expression of parental alleles during development and in the adult.' This differential modification appears to be essential for normal mammalian development since parthenogenetic embryos (whether diploid paternal or diploid maternal) do not survive to term; in diploid maternal embryos, fetal development is normal but development of the extraembryonic membranes is abnormal. In diploid paternal embryos, it is other way around. Clearly, maternal and paternal chromosomes must differ epigenetically and in such a way that different developmental programmes are followed.

Only some 100-200 genes in mammalian genomes are imprinted and these are often clustered. Human examples include the INS, H19 (D11S878E) and IGF2 (11p15.5), SNRPN (15q11-q12), WT1 (11p13), IGF2R (6q25-q27) and XIST (Xq13.2) genes. The XIST gene is essential for X-inactivation (a special case of imprinting) and is expressed exclusively from the inactive X chromosome. It gives rise to a non-coding mRNA product that controls the production of an inactivation signal which spreads along the chromosome, silencing all but a handful of genes.

A common function of imprinted genes is in the control of embryonic growth with paternally expressed genes (e.g. IGF2) tending to enhance growth rates and maternally expressed genes (e.g. H19, IGF2R) reducing them. This dichotomy has led to the proposal of the *genetic conflict hypothesis* which attempts to explain the evolution of imprinting in terms of a conflict of interest between the maternal and paternal genes of an individual. This theory might predict antagonistic coevolution between maternally and paternally derived genes for the control of fetal growth. However, contrary to the expectations of the conflict hypothesis, the rate of evolution of imprinted genes is not significantly different from that exhibited by non-imprinted genes encoding receptors.

It has been suggested that the imprinted expression of genes is usually conserved between human and mouse but whilst the maternal-specific expression of the Igf2r gene is seen in mouse, imprinting of the IGF2R gene is only apparent in a minority of the human population. Such an 'imprinting polymorphism' has also been noted for the WT1 gene in human populations.

Imprinted genes may have fewer and smaller introns than nonimprinted genes. Whether this is an adaptation to allow these genes to be transcribed rapidly or whether it is merely a property of the chromosomal region in question is however unknown.

It is present unclear if imprinting is confined to mammals. Parent-of-origin-specific effects have been claimed in the zebrafish, *Danio rerio*. It confirmed, this would be reconcilable with the observed survival of androgenetic zebrafish only if the expression of the imprinted genes were not required for development.

Repetitive Sequence Elements

Repetitive DNA comprises the bulk (>90%) of the human genome. A large number of different types of repetitive sequence element are found in the human genome and their analysis may go some way toward explaining the patterns of chromosome bands noted after the staining after the staining of metaphase chromosomes. Three main categories are now recognized: (i) a highly repetitive class including sequence families with $>10^5$ copies per haploid genome, (ii) a middle repetitive sequence class (10^2-10^5 copies), and (iii) a low repetitive class whose members possess between two and 100 copies per haploid genome. The different classes of highly repetitive satellite DNA (tandem repeats) are briefly reviewed here together with *Alu* repeats and Line (L1) elements, the two most abundant and best characterized of the interspersed middle repetitive sequence families in the human genome.

Tandem Repeats

Tandemly repetitive DNA comprises satellite DNA, minisatellite DNA and microsatellite DNA. Satellite DNA comprises the majority of heterochromatin and is clustered in tandem arrays of up to several megabases (Mb) in length. A number of different families [e.g. simple sequence (5-25 bp repeats), alphoid (169 bp and 172 bp repeat) *Sau*3A (~68 bp repeat)] have been identified which are largely confined to the centromeres. One function of satellite DNA could be to maintain regions of late replication thereby ensuring that the centromere is the last region to replicate on a chromosome.

The hypervariable minisatellite sequences (about 10^4 copies/genome) share a core consensus sequence (GGTGGGCAGARG) which is reminiscent of the *Escherichia coli* Chi element known to be a signal for generalized recombination. These minisatellites exhibit substantial copy number variability in terms of the number of constituent repeat units and are often telomeric in location.

Microsatellite DNA families are simple sequence repeats, the most common being $(A)_n/(T)_n$, $(CA)_n/(TG)_n$ and $(CT)_n/(AG)_n$ types. Minisatellites and microsatellites account for between 0.2% and 0.5% of the human genome, respectively and are widely scattered on many chromosomes. Their high copy number variability and association with a considerable number of different genes has meant that they provide a very valuable source of highly informative markers for disease analysis and diagnosis. Since their presence at specific homologous chromosomal locations is often evolutionarily conserved in primates, they may also be useful in both population genetic and evolutionary studies.

More recently, a further type of tandemly reiterated DNA sequence has been described. The 4746 bp RS447 repeat sequence, quaintly named 'megasatellite' DNA, is repeated between 50 and 70 times on chromosome 4p15. It evolutionarily conserved among mammals, polymorphic in humans and since it contains an open reading frame, it may encode a protein product.

Alu Sequences and other SINEs

The *Alu* family of short interspersed repeated elements (SINEs) is present in all primates. Up to 900,000 copies are thought to exist in the human genome (some 5% of the total DNA complement) with an average spacing of 4 kb. Most occur in noncoding DNA but some are known to be located in untranslated regions. Or even coding regions.

Alu repeats share a recognizable consensus sequence but the extent of homology to this consensus varies from 72% to 99%. Human *Alu* sequences are ~300 bp in length, are polyadenylated and consist of two related sequences each between 120 and 150 bp long separated by an A-rich region. *Alu* sequences appear to be degenerate forms of 7SL RNA (RN7SL) that have been reverse transcribed and integrated into the genome. As to a possible function for *Alu* sequences, the question still remains open. Several reports of transcription of *Alu* sequences either by RNA polymerase II or III have appeared but their transcription

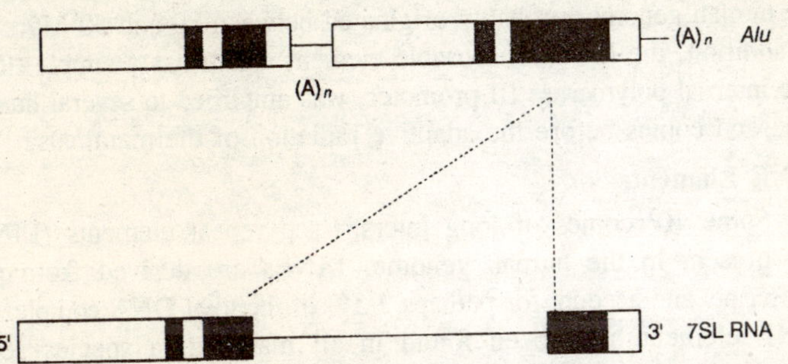

Fig. 1.7. Structure of the human Alu repeat element compared with the related 7SL RNA.

is often silenced by DNA methylation and/or nucleosome positioning. *Alu* sequences contain an internal RNA polymerase III promoter, a functional retinoic acid response element and a regulatory element that can confer positive or negative regulations of transcription upon a variety of promoters *in vitro*. The presence of the polIII promoter is important since it ensures high expression in the germline, a prerequisite for efficient transposition.

There are at least four different types of *Alu* sequence which belong to two distinct subfamilies. Some types of *Alu* sequence are human-specific and these appear to be derived from a number of different but closely related master copies or 'source genes'. The vast majority of human *Alu* sequences appear to be transcriptionally inert. Some of the human-specific subfamilies are transpositionally competent and these may also be transcriptionally active.

Alu repeats are concentrated in R bands in metaphase chromosomes whereas these repeats are under-represented in other regions (e.g. centric heterochromatin). R bands are G+C rich, replicate their DNA early in S phase and condense late in mitotic prophase. R bands also contain the bulk of active gene sequences. One consequence of the nonrandom distribution pattern of *Alu* sequences in the genome is that procedures which screen for these repeats in genomic DNA clones will preferentially locate gene sequences.

SINEs have a long evolutionary history, being present in all vertebrate classes as well as being found in molluscs. SINEs in vertebrate genomes include various types of DNA transposable element such as *Tiggers* and *mariners* which are characterized by the possessing of terminal inverted repeats and target site duplications evident as flanking direct repeats. The *Tigger* element closely resembles *pogo*, a DNA transposon in *Drosophila*. Most copies of the *mariner* family of transposable elements, of which there are probably more than 1000 in the human genome, probably originated between 80 and 50 Myrs ago. In addition, the MIR transposable element which has a simple tRNA-like internal polymerase III promoter, was amplified to several hundred thousand copies before the adaptive radiation of the mammals.

LINE Elements

Some 10^5 copies of long interspersed repeat elements (LINES) are present in the human genome. LINEs are derived from polII transcripts and account for perhaps 2-3% of the total DNA complement. LINE element have been found in all mammalian species so far examined and may be traced back to an ancestral LINE element that originated before the adaptive radiation of mammals. Human LINE elements vary in size from as little as 60 bp up to 6-7 kb; about 95% are truncated at their 5' ends but they mostly appear to contain the same 3' sequences as well as a poly(A) tail of variable length. Each individual LINE element differs from the consensus sequence by ~13% although many exhibit internal deletions and rearrangements. The majority of human LINE elements appear to have been generated within the last 30 Myrs.

LINE elements are confined to the G/Q (Giemsa/Quinacrine) bands of the euchromatin. G/Q bands are A+T rich, replicate their DNA late during the DNA synthetic period, condense early during mitosis and are relatively poor in expressed genes. LINES probably represent processed pseudogene-line copies of reverse transcripts which have been

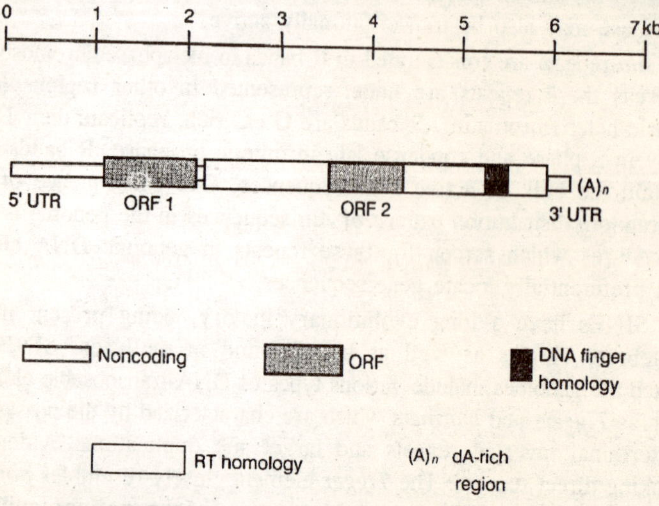

Fig. 1.8. Structure of a full-length human LINE (L1) element.

re-integrated into the genome. A full-length LINE element possesses two open-reading frames, ORF1 (1 kb) and ORF2 (4 kb); the latter possesses reverse transcriptase activity which may serve to mediate the retrotransposition not only of the LINE elements themselves but also of other retroposons such as *Alu* sequences.

LINE element transcripts have been found in undifferentiated teratocarcinoma cells suggesting that they may be expressed early on in mammalian development. A promoter at the 5' end appears to be responsible for the specific expression of LINE elements in teratocarcinoma cells. However, only a small subset of all LINE elements is capable of being transcribed.

Endogenous Retroviral Sequences and Transposons

Human transposable elements are essentially of two kinds, those that undergo transposition through a DNA intermediate (*transposons*) and those that undergo transposition through an RNA intermediate (*retroelements*). In excess of 10% of the human genome comprises integrated copies of RNA molecules including retroviruses, retroviral-like DNAs, retroposons and retrotranscripts. This represents more than 500,000 separate integration events. Nonviral retroposons include the *Alu* sequences and LINE elements discussed above. A number of endogenous retroviral or retroviral-like sequence families have been identified and characterized. These include HERV-K, RTVL-1, RTVL-H, MaLR, LTR13, and the immunoglobulin gene-related human transposon (THE1). The long terminal repeats of the HERV-K family are capable of binding human host cell nuclear proteins.

Retroviral sequences have sometimes become integrated into human genes (e.g. the endogenous retrovirus HRES-1 lies within the coding sequence of the transaldolase gene (TALDO; 11p15) and once transposed, they may even be recruited to play a role in the transcriptional regulation of a gene [e.g. as postulated for the human salivary amylase (AMYIC) gene].

GENES, MUTATIONS AND DISEASE

A central message of this volume is that the same mutational mechanisms that are responsible for disrupting the structure and function of human gene causing inherited disease are also responsible for having both created and fashioned these same genes over millions of years of evolutionary time.

In general, there are three types of mutation that give rise to an inherited disease: (i) mutations which lead to a loss of function, (ii) mutations which lead to a gain of function that is deleterious, and (iii) dominant negative mutations that adversely affect protein subunit activity or assembly. Characterized gene mutations causing genetic disease have been found to occur within coding sequences, untranslated sequences, promoter and locus control regions, in splice junctions, within introns and in polyadenylation sites. Indeed, they may interfere with any stage in the pathway of expression from gene to protein product.

Table 1.2 presents a basic classificatory system of mutation types by reference to the nature and position of the gene lesion and the stage in the expression pathway with which it interferes. Mutations are firstly divided on the basis of whether they result in the reduced synthesis of a gene product (A) or the synthesis of a structurally/ functionally abnormal gene product (B). Mutations are then secondarily divided into four categories; promoter function, gene structure, RNA processing and translation. Some gene lesions may be placed into more than one category. For example, missense mutations with drastic effects on protein structure and stability or which serve to activate an exonic cryptic splice site can fall into both categories. Similarly, a missense mutation close to an intron/exon splice junction could affect mRNA splicing efficiency as well as protein structure. The effects of specific amino acid substitutions on protein structure are the subject of several reviews.

In the context of human pathology, by far the most frequent genetic lesions in the genome are point mutations and deletions. The remainder comprise a mixed assortment of insertions, duplications, inversions, sequence amplifications and complex rearrangements. A brief review of the different types of known pathological lesion will be given.

Single Base-pair Substitutions within the Coding Region

Some 23% of point mutations are CG→TG or CG→CA transitions, representing a five-fold higher frequency for this dinucleotide than that predicted from random expectation. This is thought to be due to the hypermutability of the methylated dinucleotide CpG; spontaneous deamination of 5-methylcystosine (5mC) to thymidine in this doublet gives rise to C→T or G→A transitions depending upon the strand in which the 5mC is mutated. CpG hypermutability in inherited disease implies that the CpG sites in question are methylated in the germline and thereby rendered prone to 5 mC deamination.

The spectrum of point mutations occurring outwith CpG dinucleotides is also nonrandom. In principle, the nonrandomness of

Table 1.2. A classification of types of mutation found to cause human single gene defects through either reduced synthesis of a normal protein or normal synthesis of an abnormal protein

(a) Reduced synthesis of a normal gene product	
Defect in	
Promoter function	Binding of positive regulatory protein reduced or abolished
	Binding of negative regulatory protein increased
Gene structure	Deletions (frameshift)
	Insertions, duplications, inversions (frameshift)
RNA processing stability	Mutations in transcriptional initiation site causing failure to initiate transcription
	Splice junction mutations resulting in exon skipping and/or cryptic splice site utilization
	Activation of cryptic splice sites
	Polyadenylation/cleavage signal mutations
	Mutations in 3' untranslated region
Translation	Initiation and termination codon mutations
	Mutations in 5' untranslated region
	Nonsense mutations
(b) Synthesis of structurally/functionally abnormal gene product	
Gene structure defect resulting in	
Shortened gene product	Deletions (in-frame), nonsense mutations
Fusion genes	Deletions involving two linked genes
Elongated gene product	Insertions, duplications (in-frame)
	Termination codon mutations
Defective post-translational modification or processing, instability of protein product, impaired assembly/secretion, altered substrate/cofactor/receptor affinity.	Missense mutations

the initial mutation event, the nonrandomness of the DNA sequences under study, differences in the relative efficiency with which certain mutations are repaired, differences in phenotypic effect (and hence selection), or a bias in the clinical detection of such variants, may all play a role in determining the observed mutational spectrum.

The majority of single base-pair substitutions causing human genetic disease alter the amino acid encoded (missense mutations) but a sizeable proportion result in the introduction of a termination codon (nonsense mutations). The likelihood of clinical detection is estimated to be about three times as high for nonsense mutations as for missense mutations. Using a multidomain molecular model of the human factor IX protein, Wacey et al. (1994) have shown that the likelihood that a factor IX (F9; Xq28) gene mutation (causing hemophilia B) will come to clinical attention is a complex function of the sequence characteristic of the F9 gene, the nature of the amino acid substitution, its precise location and immediate environment within the protein molecule, and its resulting effects on the structure and function of the protein.

Single Base-pair Substitutions within Splice Sites

Splicing defects have been estimated to make up between 8% and 15% of all single base-pair substitutions causing human genetic disease. Mutations appear to occur disproportionately at the most evolutionarily conserved positions within the splice site. This is due to a detection bias resulting from the relative phenotypic severity of these lesions. Phenotypic consequences of splice site mutation include exon skipping and cryptic splice site utilization.

Single Base-pair Substitutions within Promoter Regions

A wide variety of mutations are now known which occur within gene promoter regions. These lesions disrupt the normal processes of gene activation and transcriptional initiation and serve to decrease (or rather less often, increase) the level of mRNA/protein product synthesized. What these lesions have in common is their ability to alter or abolish the binding capacity of cis-acting DNA sequence motifs for the trans-acting protein factors that normally interact with them.

Gross Gene Deletions

Gross gene deletions may be arise through a number of different recombinational mechanisms including homologous unequal recombination (occurring either between related gene sequences or repetitive elements). *Alu* sequences have been noted to flank deletion breakpoints in a considerable number of human genetic conditions and may represent hotspots for gene deletions.

Microdeletions

Deletion breakpoint junctions flanking short (<20 bp) human gene deletions are non-random both at the nucleotide and dinucleotide levels, an observation consistent with a sequence-directed mechanisms of mutagenesis. Direct repeats flanking the deleted sequence are a common finding, consistent with a model of slipped mispairing at the replication fork. Two specific types of sequence have been found at high frequency in the vicinity of short gene deletions: polypyrimidine runs at least 5 bp (YYYYY) and a 'deletion hotspot consensus sequence' (TGRRKM).

Insertions

That insertional mutagenesis might be as intrinsically non-random as point mutations and gene deletions was strongly suggested by the findings of Fearon et al. (1990), who reported 10 independent examples of DNA insertion within the same 170 bp intronic region of the DCC (18q21) gene (a locus which has been proposed to play an important role in human colorectal neoplasia). Insertional mutations involving the introduction of <10 bp DNA sequence into a gene coding region are (i) nonrandom and appear to be highly dependent upon the local DNA sequence context and (ii) may be explained by those mechanisms held to be responsible for gene deletions.

Inversions

Inversions are a highly unusual mutational mechanisms causing human genetic disease. The best known example is that found recently in the factor VIII (F8C) gene causing hemophilia A: this rearrangement occurs in about 40% of severely affected patients and recurs at high frequency. The mechanisms responsible is thought to be homologous intrachromosomal recombination between a gene (F8A) located in intron 22 of the F8C gene and two additional homologues of the F8A gene situated 500 kb upstream of the F8C gene and two additional homologues of the F8A gene situated 500 kb upstream of the F8C gene.

Expansion of Unstable Repeat Sequences

A recently recognized mutational mechanism involves the instability of certain specific trinucleotide repeat sequences. This mechanism was first reported s a cause of the fragile X mental retardation syndrome. The brain-expressed FMR1 gene responsible was found to contain an unusual (CGG)n repeat which exhibited copy number variation of between 6 and 54 in normal healthy controls, between 52 and >200 in phenotypically normal transmitting males (the 'premutation') and between 300 and >1000 in affected males (the 'full mutation'). Expansion of premutations to full mutations occurs only during female meiotic transmission whilst the probability of repeat expansion correlates with repeat copy number, consistent with a mechanism of slipped mispairing during replication. Expansion of a sequence can thus itself lead to further expansion correlates with repeat copy number, consistent with a mechanism of slipped mispairing during replication. Expansion of a sequence can thus itself lead to further expansion, a process termed 'dynamic mutation' by Richards et al. (1992). The discovery of this novel mutational mechanism soon led to the recognition that the expansion of unstable repeats is responsible for a number of other human inherited diseases, almost all·neuromuscular.

The triplet repeats (GGG)n and (CAG)n are very abundant in the human genome. A considerable number of human genes, both uniquitously expressed and tissue-specific, have now been identified as containing such triplet repeats. Many of these repeats are highly polymorphic and may thus represent examples of more subtle triplet repeat expansions. Whether specific polymorphic alleles are associated with any particular phenotype, however, remains to be seen.

This chapter has attempted to provide the reader with a brief introduction to what is known of structure-function relationships in the human genome. Specific DNA sequences have clearly evolved for different cellular functions. Some sequences represent gene coding regions that contain the genetic information necessary to direct the synthesis of proteins. Other sequences in gene promoter or untranslated regions serve to direct appropriate gene expression or are involved in mRNA processing, stability and nucleocytoplasmic transport. Some highly repetitive sequences may play an important role in chromosome architecture whereas others may almost be commensally parasitic in that they cohabit in the genome having increased in copy number under their own mutational momentum. Some sequences are modified by DNA methylation which itself may influence function. Other sequences, by their very nature, are more mutable than others and may be capable of rapid change. Whatever their cellular role, many DNA sequences are able to interact with proteins or mRNA molecules in order to effect their function. It follows that the imminent analysis of the human genome sequence should reveal the existence of new sequence codes (and perhaps codes within codes) to add to that first elucidated in the early 1960s.

2

Organization of Genome

There are several aspects of genome organization and evolution. The first is genome size, which varies enormously among organisms. How did this variation occur and what mechanisms can increase or decrease genome size to produce such variation? The second topic is the genetic information included within genomes. Do genomes mostly consist of genic DNA? Or, is the genome made of mostly nongenic sequences? Are there many repetitive sequences in the genome, and if so, what is the pattern of chromosomal distribution of the repeated sequences? Does the nongenic fraction have a function, or is it merely "junk"? The third topic concerns the variation in nucleotide composition over different parts of the genome. Is there heterogeneity in composition among different regions of the genome? What mechanisms can give rise to localized differences in nucleotide composition? Finally, why are there numerous spliceosomal introns in the genome of higher eukaryotes, but no such introns have ever been found in prokaryotic gene? What is the origin of introns and what is their evolutionary significance?

Genome Size Variation and the C-value Paradox

The genome size of an organism is defined as the amount of DNA in the haploid genomic set, such as that in the sperm nucleus. It is also commonly called the C value, where C stands for "constant" or "characteristic" to denote the fact that the size of the haploid genome is fairly constant within anyone species. In contrast, C values vary widely among species in both prokaryotes and eukaryotes. Large genomes such as the nuclear genomes of eukaryotes are usually measured in picogram (pg) of DNA (1 pg = 10^{12}g), whereas small genomes such as the genomes of bacteria and viruses are more commonly measured in base pairs (bp) or kilobase pairs (Kb) or in daltons. For convenience, we shall use mainly the unit of one million base pairs (Mb) because most genome sizes are larger than 1 Mb and because this unit is now commonly used in the human genome project. The conversion factors are 1 pg $\approx 0.98 \times 10^9$ bp = 980 Mb and 1 dalton $\approx 1.62 \times 10^{-3}$ bp.

Genome Sizes of Bacteria

Bacterial genome sizes vary over a 20-fold range, from about 0.6 Mb in some obligatory intracellular parasites, to more than 10 Mb in several cyanobacterial species. The smallest prokaryotes, the mycoplasmas, have a genome of ~0.6 Mb, which consists of about 470 protein-coding genes including some 50 ribosomal proteins, one or two sets of rRNA genes (5S, 16S, and 23S), and about 33 tRNA genes. Therefore, the genome of mycoplasma contains about 500 genes, which is still smaller but probably close to the minimum number for supporting a free-living cell or life organism. The number of genes in *Escherichia coli* is probably between 3000 and 6000. The numbers of genes in other bacteria range very roughly from about 500 to 8000. In other words, among characterized bacterial species, the variation in the number of genes is the same as the variation in C values. Therefore, bacteria do not seem to contain large quantities of nongenic DNA.

Table 2.1. Range of C values in bacteria

	Range in genome size (kb)	*Ratio (highest/lowest)*
Eubacteria	580-13,200	20
Mycoplasmas	580-1,800	3
Gram negative	650-7,800	12
Gram positive	1,600-11,600	7
Cyanobacteria	3,100-13,200	4
Archaebacteria	1,600-4,100	3

The genomes of bacteria can be divided into three fractions: (1) chromosomal DNA, (2) DNA that originated in plasmids, and (3) transposable elements. The chromosomal fraction contains protein-coding genes required for growth and metabolic functions (70-80%), spacers and various signals (20-30%), RNA-specifying genes (~1%), and a number of short repeated sequences, usually on the order of a few dozens of base pairs in length. Bacteria carry many plasmids as extrachromosomal genetic elements. In some instances, however, genes derived from plasmids are found in the bacterial chromosome. Transposable elements are common components of some bacterial genomes. For example, the chromosome of *E. coli* contains 1-10 copies of at least six different types of insertion sequences. The non-chromosomal fraction of the genome (including insertion sequences and plasmid-derived genes in the chromosome) seems to be one order of magnitude small than the chromosomal fraction.

The distribution of gene sizes in bacteria is discontinuous, showing major peaks with modal values of about 0.8, 1.6, 4.0 Mb, and several minor peaks at 7.2 and 8.0 Mb. This distribution has led to the suggestion that the larger genomes evolved from smaller genomes by successive genome duplications. Since there seems to be no notable relationship between genome size and bacterial phylogeny, it has been suggested that genome duplications have occurred frequently in the evolution of bacterial lineages.

Using a tentative phylogeny of bacteria based on comparison of rRNA sequences, Herdman (1985) was able to relate changes in genome size to phylogenetic history. The results indicate that genome duplication occurred independently in different bacterial lineages. Interestingly, many of the duplications seem to have occurred coincidentally in several bacterial lineages at a rather specific time in evolution, i.e., soon after the appearance of oxygen in the atmosphere, approximately 1.8 billion years ago.

In summary, the distribution of genome sizes in bacteria can be explained by a combination of several processes: (1) genome duplication during the independent evolution of a respiratory metabolism in several lines, (2) subsequent whole genome duplications occurring independently in many lineages, (3) small-scale deletions and insertions, (4) duplicative transposition, (5) horizontal transfer of genes derived mainly from plasmids, and (6) loss of massive chunks of DNA in many parasitic lines.

Genome Sizes of Eukaryotes

C values in eukaryotes are usually much larger than in prokaryotes; for example, the genome size of humans (~3400 Mb) is about 1000 times that of *E. coli* (~4.7 Mb). However, there are exceptions. For instance, the yeast *Saccharomyces cerevisiae* has a smaller genome than many Gram-positive bacteria and most cyanobacteria. Since the eukaryotic genome has multiple origins of replication, while most prokaryotes seem to have only one, the eukaryotes are able to replicate much larger amounts of DNA in an amount of time not much larger than that required for a much smaller prokaryotic genome.

Table 2.2. Range of C values in various eukaryotic groups of organisms

	Range in genome size (kb)	*Ratio (highest/lowest)*
Protists	23,500-686,000,000	29,191
Euglenozoa	98,000-2,350,000	24
Ciliophora	23,500-8,620,000	367
Sarcodina	35,300-686,000,000	19,433
Fungi	8,800-1,470,000	167
Animals	49,000-139,000,000	2,837
Sponges	49,000-53,900	1
Annelids	882,000-5,190,000	6
Molluscs	421,000-5,290,000	13
Crustaceans	686,000-22,100,000	32
Insects	98,000-7,350,000	75
Echinoderms	529,000-3,230,000	6
Agnathes	637,000-2,790,000	4
Sharks and rays	1,470,000-15,800,000	11
Bony fishes	382,000-139,000,000	364
Amphibians	931,000-84,300,000	91
Reptiles	1,230,000-5,340,000	4
Birds	1,670,000-2,250,000	1
Mammals	1,420,000-5,680,000	4
Plants	50,000-307,000,000	6,140
Algae	80,000-30,000,000	375
Pteridophytes	98,000-307,000,000	3,133
Gymnosperms	4,120,000-76,900,000	17
Angiosperms	50,000-125,000,000	2,500

The variation in C values in eukaryotes is much larger than that in bacteria, from ~9 Mb to ~690,000 Mb, approximately a 80,000-fold range. Unicellular protists, particularly sarcodine amoebae, show the greatest variation in C values. The three amniote classes (mammals, birds, and reptiles), on the contrary, are exceptional among eukaryotes in their small variation in genome size (up to fourfold). Other classes, for which a substantial body of C-value data exist, show variation of a least 100-fold.

Table 2.3. C values from eukaryotic organisms ranked by size

Species	*C value (kb)*
Navicola pelliculosa (diatom)	35,000
Drosophila melanogaster (fruitfly)	180,000
Paramecium aurelia (ciliate)	190,000
Gallus domesticus (chicken)	1,200,000
Erysiphe cichoracearum (fungus)	1,500,000
Cyprinus carpio (carp)	1,700,000
Lampreta planeri (lamprey)	1,900,000
Boa constrictor (snake)	2,100,000
Parascaris equorum (roundworm)	2,500,000
Carcarias obscurus (shark)	2,700,000
Rattus norvegicus (rat)	2,900,000
Xenopus laevis (toad)	3,100,000
***Homo sapiens* (human)**	**3,400,000**
Nicotiana tabaccum (tobacco)	3,800,000
Paramecium caudatum (ciliate)	8,600,000
Schistocerca gregaria (locust)	9,300,000
Allium cepa (onion)	18,000,000
Coscinodiscus asteromphalus (diatom)	25,000,000
Lilium formosanum (lily)	36,000,000
Amphiuma means (newt)	84,000,000
Pinus resinosa (pine)	68,000,000
Protopterus aethiopicus (lungfish)	140,000,000
Ophioglossum petiolatum (fern)	160,000,000
Amoeba proteus (amoeba)	290,000,000
Amoeba dubia (amoeba)	670,000,000

Interestingly, the huge interspecific variation in genome sizes among eukaryotes seems to bear no clear relationship to either organismic complexity or the likely number of genes encoded by the organism. For example, several unicellular protozoans possess much more DNA than mammals, which are presumably more complex. Moreover, organisms that seem similar in complexity (e.g., flies and locusts, onion and lily, *Paramecium aurelia* and *P. caudatum*) possess vastly different C values. This lack of correspondence between C values and the presumed amount of genetic information contained within genomes has become known in the literature as the *C-value paradox*. The C-value paradox is also evident in comparisons of sibling species (i.e., species that are so similar to each other morphologically as to be indistinguishable phenotypically). In protists, bony fishes, amphibia, and flowering plants, many sibling species differ greatly in their C values, in spite of the fact that, by the definition of sibling species, no difference in organismic complexity exists. Since we cannot assume that a species possesses less DNA that the amount required for specifying its vital functions, we have to explain why so many species contain vast excesses of DNA.

The first question to be clarified in whether or not a correlation exists between genome size and the number of genes. In other words, are the differences in genome sizes attributable to genic DNA or to nongenic DNA? Eukaryotes show approximately 20-fold variation in the number of protein-coding genes, from about 6000 in yeast to roughly 60,000-80,000 in mammals. This 20-fold variation is insufficient to explain the 80,000-fold variation in nuclear DNA content. One way to compare the gene numbers in two genomes is to analyze the polysomal RNA complexity in the same tissue of the two organisms; the RNA complexity refers to the total length of different species of RNA. Such data are available in several vertebrates. In *Xenopus laevis* and *Triturus cristatus* egg RNA complexities are of the order of 30-40 million nucleotides (Mnt). Thus, the two amphibians have an equivalent RNA transcribing capacity, despite the eightfold difference in genome size. As another example, the chicken has a genome size one-third of that of the mouse, yet the polysomal RNA complexities in chicken and mouse livers are virtually the same. Therefore, the large interspecific variation in genome size among organisms appears to be mostly due to nongenic DNA rather than genic DNA.

A positive correlation between the degree of repetition of several RNA-specifying genes and genome size has been found. Similarly, a correlation exists between genome size and the number of copies of such regulatory sequences as telomeres, centromeres, and replicator genes, which are required for chromosome replication, segregation, and recombination during meiosis and mitosis. However, all these genes constitute only a minute fraction of the genome, such that the variation in the number of RNA-specifying genes and regulatory sequences cannot explain the variation in genome size.

In summary, we are left with the nongenic DNA fraction as the sole culprit for the C-value paradox. In other words, a substantial portion of the eukaryotic genome consists of DNA that does not contain genetic information. It has been estimated that the amount of nongenic DNA per genome varies in eukaryotes from about 3 Mb to over 100,000 Mb (a 100,000 fold-range) and constitutes anything from less than 30% to almost 100% of the genome.

Repetitive Structure of the Eukaryotic Genome

The eukaryotic genome is characterized by the presence of repetitive DNA consisting of nucleotide sequences of various lengths and compositions that occur from a few times to millions of times in the genome either in tandem or in a dispersed fashion. Segments of DNA that are not repeated are referred to as *single-copy DNA* (*scDNA*) or *unique DNA*. The proportion of the genome taken up by repetitive sequences varies widely among taxa. In yeast, this proportion amounts to about 20%. In animals, the proportion ranges from a few percent as in *Chironomus tentans*, an arthropod, to close to 90% (i.e., only 10% scDNA) in *Necturus masculosus*, an amphibian. In plants, the proportion of repetitive DNA can exceed 80%.

Classical studies of the kinetic DNA reassociation by Britten and Kohne (1968) showed that the genome of eukaryotes can roughly be divided into four fractions: *foldback DNA*, *highly repetitive DNA*, *middle-repetitive DNA*, and single-copy DNA. Foldback DNA consists of palindromic DNA sequences that can form hairpin double-stranded

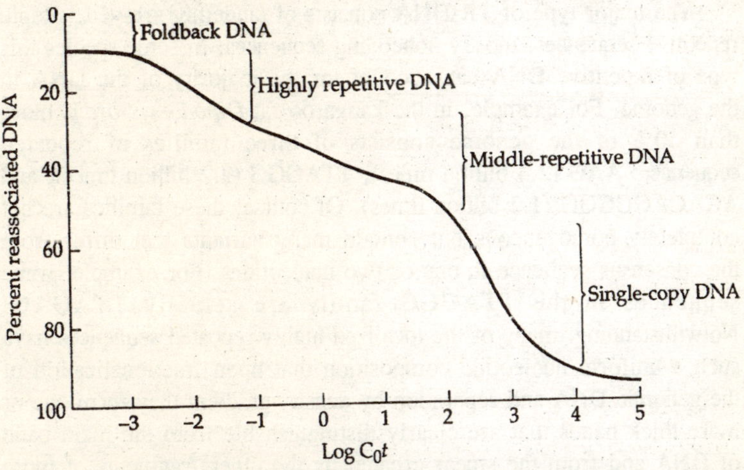

Fig. 2.1. A reassociation profile of mammalian DNA.

structures as soon as the denatured DNA is allowed to renature. The highly repetitive fraction is made up of short sequences, from a few to hundreds of nucleotides long, which are repeated thousands or even millions of times. The middle-repetitive fraction consists of much longer sequences, hundreds of thousands of base pairs on the average, which appear in the genome from dozens to thousands of times.

On the basis of the pattern of dispersion of repeats, the repetitive fractions can be classified into two types of repeated families: localized and dispersed.

Localized Repetitive Sequences

Localized repetitive sequences usually occur as tandem arrays and, when this is the case, they are called *tandem repetitive DNA (TR-DNA)*. One type of TR-DNA consists of multigene families with members arranged in tandem; however, some multigene families have members dispersed to various chromosomal locations. The repeat unit in a gene family generally includes the entire sequences of the gene (exons, introns, and untranscribed flanking sequences) and spacer regions, and thus can be several kilobases or even larger in length. Well-known examples of tandemly arrayed multigene families are rRNA and tRNA genes in mammals. For example, the human genome contains approximately 350 rRNA genes and 1300 tRNA genes; the rRNA genes are located in the nucleolar organizers of the five acrocentric chromosomes (13, 14, 15, 21, and 22). The β-globin cluster in humans represents an example of a low copy number multigene family.

The major type of TR-DNA consists of tandemly arrayed, simple repetitive sequences, mostly noncoding sequences. In some species this type of repetitive DNA can account for the majority of the DNA in the genome. For example, in the kangaroo rat *Dipodomys ordii*, more than 50% of the genome consists of three families of repeated sequences: AAG (2.4 billion times), TTAGGG (2.2 billion times), and ACACAGCGGG (1.2 billion times). Of course, these families are not completely homogeneous but contain many variants that differ from the consensus sequence in one or two nucleotides. For example, some sequences in the TTAGGG family are actually TTAGAG. Notwithstanding, many of the localized highly repeated sequences have such a uniform nucleotide composition that upon fractionalization of the genomic DNA and separation by density gradient they form one or more thick bands that are clearly distinguishable from the main band of DNA and from the smear created by the other fragments of more heterogeneous composition. These bands are called *satellite DNA*.

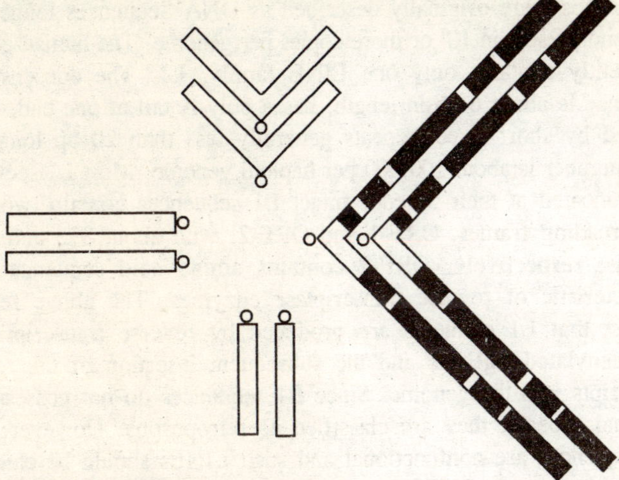

Fig. 2.2. Highly repetitive DNA sequences (black areas) in Drosophila nasutoides are localized on a single chromosome.

In some species members of a satellite DNA are found on more than one chromosome, while in others they are restricted to a particular chromosomal location. For example, alpha satellite or alphoid DNA, the most abundant family of satellite DNA in the human genome, is found at the centromeres of all chromosomes. In contrast, 60% of the genome of *Drosophila nasutoides* consists of satellite DNA, and all of its is localized on one of the four chromosomes, which seems to contain little else. In general, satellite DNA is located inside heterochromatic or centromeric regions. Other types of simple tandem repetitive sequences will be discussed in detail in the next section.

Dispersed Repetitive Sequences

The second class of repetitive DNA consists of sequences that are dispersed throughout the genome. Single copies of dispersed repetitive sequences are found in introns, flanking regions of genes, intergenic regions, and nongenic DNA. There are two major kinds of dispersed highly repetitive sequences: short interspersed repeats (SINEs) and long interspersed repeats (LINEs). In addition, there are moderately repetitive sequences, which are usually transposable elements or processed pseudogenes and are usually dispersed over the genome.

SINEs are retropseudogenes, i.e., pseudogenes derived from POL III RNA transcripts. As has been discussed, they are abundant in mammalian genomes. For example, there are more than 500,000 Alu repeats in the human genome.

LINEs were originally described as DNA Sequences longer than 5 kb and present in 10^4 or more copies per genome. The human genome apparently contains only one LINE family, L1. The consensus L1 sequence is about 6 kb in length, has a poly-A tail at one end, and is flanked by short direct repeats generally less than 20 bp long. The copy number is about 100,000 per haploid genome. Most L1 sequences are truncated at their 5' end. Intact L1 sequences contain two large open reading frames, ORF-1 and ORF-2, with about 375 and 1,300 codons, respectively. ORF-2 contains amino acid sequence motif characteristic of reverse transcriptase enzymes. The above features suggest that L1 elements are produced by reverse transcription of polyadenylated mRNAs and the subsequent insertion of the reverse transcripts into the genome. Since L1 sequences do not possess long terminal repeats, they are classified as retroposons. However, most L1 sequences are nonfunctional and such LINEs should be classified as retrosequences. LINEs homologous to the human L1 family have been found in all mammals, including marsupials. Nonmammalian elements related to L1 include the I, F, G, and D factors in *Drosophila* species, *Ingi* in *Trypanosoma brucei*, and R2 in *Bombyx mori*.

Most L1 sequences are truncated, containing no intact reading frames. Such defective elements may not be able to transpose and were found to evolve much more rapidly than intact elements, probably because they are no longer subjected to functional constraints. The act that most L1 sequences are defective implies that the propagation of L1 elements within the genome depends on only a small number of source elements. For these reasons, the degree of L1 sequence divergence among species is much greater than the degree of divergence among conspecific L1 copies. For example, L1 sequences from mice and humans differ from each other on average by about 30%, whereas the average sequence divergence among L1 elements in mice is only 4%. Hardies et al. (1986) estimates that in rodents more than half of the L1 elements are no more than 3 million years old.

Genomic Location of Transcribed Genes

RNA-DNA hybridization experiments have shown that only a very small fraction of the transcribed genes are located within the repeated fractions of the genome. Most transcribed genes resides within the unique DNA fraction, most of which is in euchromatic regions. Even in the unique fraction, most sequences are not transcribed. Indeed, it has been estimated that only about 3% of the non-repetitive DNA sequences in humans are transcribed. These studies constitute further support for the view that most of the eukaryotic genome is devoid of genetic information.

Tandem Repetitive Sequences

Tandem repetitive DNA (TR-DNA) has attracted much attention because many TR-DNA loci are highly variable among individuals in a population and so are very useful genetic markers in biology and medicine. This line of research started with Wyman and White's (1980) fortuitous observation of locus highly polymorphic for restriction length variation and the finding by Bell et al. (1982) that the hypervariable locus near the insulin gene consisted of tandem repeats, with multiallelic variation arising through variation in repeat copy number. Subsequently, probes constructed from a short tandem-repeat region in an intron of the human myoglobin gene were shown by Southern blot hybridization to detect an astonishingly complex variable array of DNA fragments, not just from humans but also from several other animals species that happened to be on the Southern blot. This was the beginning of the so-called DNA fingerprinting study.

Classification of Tandem Repetitive DNA

Table 2.4 shows a classification of simple tandem repetitive sequences, according to the size of the repeat unit, the number of repeat units per array, and the genomic location of the tandem arrays. The first class, satellite DNA, has already been discussed in the preceding section; satellite DNA has rarely been used in fingerprinting because it is too large for PCR amplification. The rest of this section concerns the other four classes. It should be noted that there is presently no consensus in the classification of simple tandem repeats and the terminology used varies among authors.

Table 2.4. Five classes of simple tandem repetitive sequences

Class	*Examples*	*Repeat size (base pairs)*	*Array size (repeat units)*	*Genomic distribution*
Satellites	Alpha, Beta	2-100s	>1000s	Centromeric, heterochromatic
Minisatellites	Apo B, D1S7	9-100	10-100	Dispersed, subtelomeric
Short tandem repeats	FM-1, MYCN	3-5	10-100	Dispersed
Microsatellites	Apo CII, PLAT	1-2	10-100	Dispersed
Alu tail array	IN20-Rep HMG-CoA	1-5	10-40	Dispersed

The term *minisatellite* was originally coined and is still used by Jeffreys and colleagues to denote any tandem repeat region that is much shorter than but has a structural similarity to classical satellite DNA. However, it is used here and by other authors in a more restricted sense, i.e., as a tandem array of repeats with a unit length 9 bp or longer, while tandem arrays of repeats with a shorter unit length are classified into three other classes. The advantage of having more classes of TR-DNA in terms of repeat length is that the primary mechanism for generating repeated sequences may be different from different length classes. The term *short tandem repeats* (STRs) refers to tandem repeats with repeat units 3-5 bp long. It has been estimated that there are 300,000 trinucleotide and tetranucleotide STRs in the human genome or, on average, one array in every 10 kb of genomic DNA. *Microsatellites* are TR-DNA arrays that have repeat units of 1-2 bp; some authors also include STRs in the class of microsatellites. The most common human microsatellites are dinucleotide arrays of $(CA)_n$, which means n repeats of CA. There are approximately 50,000 $(CA)_n$ arrays in the human genome or about one array every 30 kb. Alu tails, the last class, are tandem arrays of repeat units 1-5 bp in size that are present immediately 3' of Alu elements. The most common Alu tail arrays are $(A)_n$ and $(AAAN)_n$ where N=G, C or T. Interestingly, tandem arrays with repeat unit length 6-8 be pare rarely observed and so are not included.

A tandem repeat region that shows variation in number of repeats among chromosomes in a population is commonly referred to as a VNTR locus, where VNTR stands for *variable number of tandem repeat*. Since most tandem repeat regions are variable, VNTRs include virtually all the last four classes of loci; satellite DNA loci are not included because variation in size between individuals is difficult to type. The most prominent feature that distinguishes VNTR loci from classical gene loci is that they are usually highly polymorphic, having a large number of alleles per locus and a high heterozygosity. Indeed, the heterozygosity can be close to 1. The high polymorphism appears to be largely due to exceptionally high rates of mutation; e.g., the mutation rate per gamete at a minisatellite locus can be a few percent per generation, in contrast to 10^{-5} per generation or lower at classical gene loci.

Currently VNTRs are heavily used in biology, medicine, and forensic science. Since a large number of alleles usually exist at a VNTR locus, a collection of several VNTR loci can lead to a huge number of possible combinations or DNA profiles, making VNTR loci extremely powerful for identity analysis and forensic use. For example, in the investigation of a crime where a biological specimen has been obtained from the crime scene, DNA from the specimen and from the suspect are typed for size at several unlinked VNTR loci. If the profiles do not match, the suspect is excluded. If they match, the probability of this match occurring in the population by chance alone can be calculated from the frequencies of alleles in the population. This probability is usually extremely low because of high polymorphism at VNTR loci. In the United States, however, the application of VNTRs in crime courts is presently still controversial, largely because of disagreement on how to compute the frequency of a specific profile, e.g., whether it should be computed on the basis of the ethnic group the crime suspect belongs to or on the basis of the general population. The high polymorphism and the ease of typing make VNTR loci also very useful for genetic mapping, studies of population structure, detecting loss of heterozygosity in tumor initiation and progression, ethological studies, and disease association studies.

The interest in VNTR loci has been further stimulated by the recent finding that mutation at some VNTR loci can cause a severe disease. For example, the following four genetic diseases have been associated with unstable VNTR loci: Kennedy's disease, fragile X mental retardation syndrome, myotonic dystrophy and Huntington's disease. All four diseases and at least six others are caused by amplification of trinucleotide tandem arrays, i.e., an allele becomes disease-causing when it has more repeats than some threshold number. For example, in the case of fragile X the threshold number is about 200; the normal gene contains 6 to 54 copies of the CGG repeat and a *premutation* allele has 50 to 200 repeat units. Surprisingly, this VNTR locus is in the 5' untranslated region of the fragile X gene and in the case of myotonic muscular dystrophy the VNTR locus is in the 3' untranslated region rather than in the coding region of a protein kinase gene.

Mutational Mechanisms

Current models of size change at VNTR loci all involve mispairing of repeats but differ in the timing of the DNA mispairing event: replication, mitosis, and meiosis. *Slipped-strand mispairing* (SSM) or *replication slippage* occurs during DNA replication; *unequal sister -chromatid exchange* (USCE) occurs during mitosis and, less frequently, during meiosis; and unequal genetic recombination or crossing-over occurs during meiosis. SSM and USCE are intrachromosomal phenomena because they involve exchange between two copies of the same chromosome, whereas meiotic unequal crossing-over is an interchromosomal phenomenon because it involves two different chromosomes. Meiotic unequal crossing-over causes an exchange of flanking markers, whereas SSM and USCE do not. This property can be used to detect meiotic unequal crossing-over events. Wolff et al. (1989) found that only 1 in 12 size-change events at the D1S7 minisatellite locus had a recombination between two flanking markers, and Vergnaud et al. (1991) observed no exchange of flanking markers in 52 mutation events at the CEB1 locus. Another way to study the mechanism of mutation is to determine from which parental allele(s) a mutant allele is derived. Jeffreys et al. (1990, 1991) developed the minisatellite variant repeat (MVR) mapping technique, which utilizes sequences differences between repeat units to assay the interdispersion pattern of different repeat units along an allele. Using this technique, Jeffreys et al. (1990) determined that each of the 38 mutant alleles isolated from sperm and the 37 mutant alleles from blood at the D1S8 minisatellite locus could be traced back to a single parental allele; thus none of these mutant alleles had arisen by recombination between homologous chromosomes. In a subsequent study, Jeffreys et al. (1991) found that 2 of the 7 additional mutant alleles at the D1S8 locus had arisen by recombination events between parental alleles. The above data taken together suggest that meiotic unequal crossing-over is a minor mechanism of mutation at minisatellite loci or, in other words, size change at minisatellite loci occurs mainly at mitosis. This conclusion is also supported by the observation that only 1 of the 52 mutations at the CEB1 locus was of maternal origin. (As explained already, in the formation of mammalian germ cells, sperm undergo many more rounds of mitotic division than do eggs but the same number of meiotic divisions. Thus, a high proportion of mutants of paternal origin suggests that mutation occurs mainly at mitosis.) Unfortunately, the above techniques cannot distinguish between USCE and SSM. However, USCE events tend to be larger in size than SSM events because USCE is less constrained in misalignment of repeats than is SSM. The average size of change for the 52 mutations at the CEB1 locus was 250 bp or 9% of the size of the parental allele. Moreover, the average size change for the 32 mutations at the D1S7 locus studied by Jeffreys et al. (1988) was 34 repeat units or 5% of the parental allele. The relatively large size of these mutation events suggests that USCE is the primary mechanism of mutation at these minisatellite loci. However, recent data suggest that minisatellites do not primarily mutate by processes such as slippage and unequal crossover intrinsic to the tandem repeat array. Instead, germline repeat instability is largely regulated by cis-acting elements near the arrays and involves unexpectedly complex processes of gene conversion.

Limited data on the size change and parental origin of mutation events at microsatellite and STR loci are also available. Of the 7 mutation events where the parental origin was determined, 6 were of parental origin and only 1 was maternal. The prevalence of paternal origin suggests that size mutation at these loci occurs mainly at mitosis. Of the 9 mutations 6 were changes of 1 repeat unit, 2 were changes of 2 repeat units and 1 was a change of 3 repeat units. The small size of the mutation events suggests SSM as the primary mechanism of mutation for microsatellite and STR loci. Thus, while USCE seems to be the primary mechanism of mutation at minisatellite loci, SSM seems to be the primary mechanism of mutation at microsatellite and STR loci. However, since this conclusion is based on limited data, it needs to be substantiated by further studies.

Population Genetics of VNTRs

Another way to study the mutational mechanisms is to use the theory of population genetics. The change in the number of repeats in an allele can be assumed to follow the stepwise mutation model of Ohta and Kimura (1973). The simplest model is the one-step model, which assumes that each mutation increases or decreases the number of repeats in the allele by one unit. The distribution of allele frequencies under this model can be compared with data. One interesting observation from data is that the distribution of allele frequencies versus the allele size (number of repeats) is often multimodal. One might wonder whether the multimodality arises from population subdivision, migration, or selection. However, Valdes et. al (1993) showed that a multimodal distribution often arises in a random mating population under the one-step model without migration or selection. Actually, it is not difficult to imagine in terms of gene genealogy how a multimodal distribution can arise from stepwise mutation. For example, if the genealogy of the sequences in the sample, where the sequences are clustered into two groups of closely related sequences, then it is likely to have a bimodal distribution because the sequences on each side of the root would have similar numbers of repeats but would differ from those on the other side in repeat number. Valdes et al (1993) also compared the observed variances of the

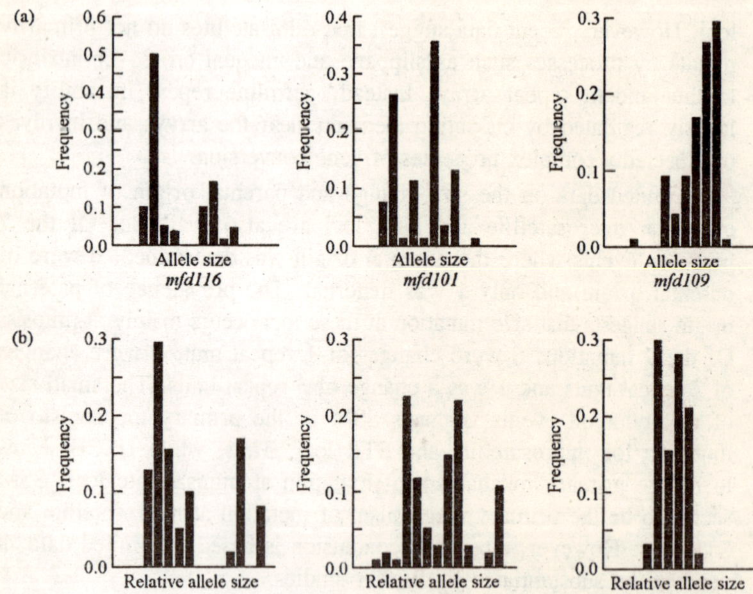

Fig. 2.3. (a) Frequency distributions of alleles at three microsatellite loci. (b) Frequency distributions of alleles for three independent replicates in a computer simulation.

distribution of allele size at 108 microsatellite loci (arrays of dinucleotide repeats) with simulation results, and concluded that the observed distributions are consistent with the results of the one-step model.

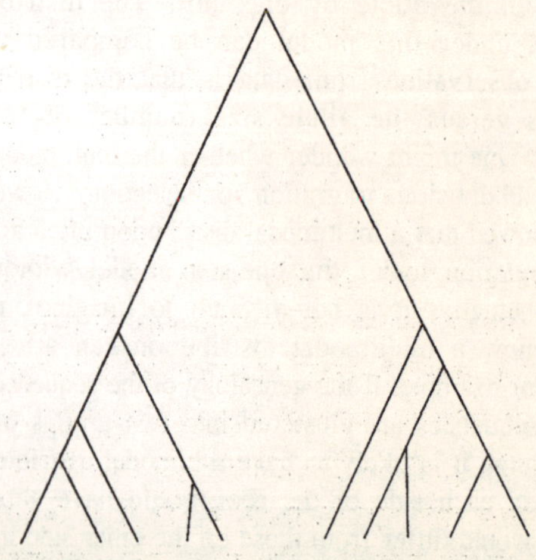

Fig. 2.4. A genealogical tree for a sample of DNA sequences.

Shriver et al. (1993) concluded a more detailed study by investigating four summary measures of allele frequency distributions—number of alleles, number of modes, range in allele size, and heterozygosity—by using computer simulations of the one-step model. The VNTR loci analyzed were classified into microsatellites (1-2 bp repeat unit), STRs (3-5 bp repeat unit) and minisatellites (15-70 bp repeat unit). They chose data from the literature that were obtained from large homogeneous populations to avoided the effects of population substructure and recent decreases in population size. For the 12 STR loci examined there were no significant deviations for any summary measure predictions, and so it appeared that these loci have a mutational mechanism very much like the one-step model. Of the hypothesized mechanisms of VNTR mutation, replication slippage is most like the stepwise mutation model and so is the most likely mechanism of mutation at STR loci. By comparison, 35% of the 31 microsatellite loci examined had significant deviations from the simulation expectations for the number of alleles or their size range and 19% of the loci were borderline significant for deviation from the expected number of models. Thus, it appeared that the mutational mechanism of microsatellite loci is close to but it not exactly a one-step type mechanism. A plausible explanation for the small, but significant, excess in number of alleles and allelic range for this class of loci is that the mutational mechanisms is stepwise with a low frequency of multistep mutations in addition to single-step mutations. Finally, only 3 or 11 minisatellite loci showed no significant differences between the observed summary measures and their simulation results. Multiple mechanisms may be involved in the mutation of minisatellite loci.

The infinite-allele model has also been examined by comparing its predictions with data or with simulation results. In contrast to the stepwise model, this model assumes that each mutation produces a new allele. The general conclusion is that this model fits data less well than does the stepwise mutation model, particularly for STR and microsatellite data. Harding et al's (1993) simulation results based on unequal sister-chromatin exchange also show the superiority of the stepwise model over the infinite-allele model.

Many questions concerning the population genetics of VNTRs remain unsolved. For example, the effects of migration, change in population size, and natural selection on the distribution of allele frequencies have not been well studied. A better understanding of the population genetics of VNTRs will strengthen their applications, e.g., in forensic science and in making biological inferences such as estimation of effective population size and the relationships among populations.

Mechanisms for Increasing Genome Size

Before attempting to explain the maintenance of the vast amount of nongenic DNA in the genome of eukaryotes, we consider possible mechanisms for increasing the size of a genome. The putative mechanisms should be able to explain not only the existence of nongenic DNA *per se*, but also its repetitiveness and particular genomic distribution.

We distinguish between two types of genome increase: (1) global increases, in which the entire genome or a substantial part of it, such as a chromosome, is duplicated, and (2) regional increases, in which a particular sequence is multiplied to generate repetitive DNA. The latter mechanisms include transposition, replication slippage, unequal sister-chromatid exchange, interchromosomal unequal crossing-over, and DNA amplification. Transposition produces dispersed repetitive sequences, while the other mechanisms produce localized repetitive sequence.

Genome Duplication

Genome duplication occurs as a consequence of lack of disjunction between all the daughter chromosomes following DNA replication. Since it immediately doubles the size of a genome, it would be the most effective mechanism for increasing genome size provided that it does not cause deleterious effects. Indeed, if genome duplication were the sole means for genome enlargement, it would require only 10 rounds of genome duplication to explain the thousandfold increase in genome size from *E. coli* to mammals. Of course, the assumption of no deleterious effects is often not true, and so the question is: How much has genome duplication really contributed to the increase in genome size in the evolution from bacteria to mammals or during the evolution of a group of organisms?

In plants, polyploidy is a widespread phenomenon. For example, in monocotyledons genome sizes exhibit a polymodal distribution with peaks at 0.60, 1.18, 2.16, 4.51, and 8.53×10^9 bp, a distribution suggesting a series of genome duplication. Numerous studies have been made to understand the prevalence of polyploidy in plants. Artificially produced autopolyploids are generally inferior to their diploid progenitors. The inferiority is expressed in lower fertility and also often in lowered ability to compete with diploids in artificially controlled experiments and under more or less natural conditions. The inferiority seems to be mainly due to physiological effects such as genetic imbalances and prolongation of cell-division time and partly due to irregularities of chromosomal segregation. These experimental results and the fact that the performance of induced polyploids as agricultural crops has consistently fallen short of expectations have led the majority of workers to believe that genome multiplication *per se* is selectively disadvantageous. The widespread success of polyploidy in plants has therefore been attributed to hybridization between individuals from different populations such as ecotypes, races, or species; hybridization increases genetic diversity and may therefore enable the polyploids to compete with their progenitors or to colonize new habitats.

Polyploidy is extremely rare in bisexually reproducing animal groups. Muller (1925) proposed that this is because in bisexual animals the two sexes are differentiated by means of a process involving the diploid mechanism of segregation and combination, and polyploidy invariably disturbs this process. [Muller's explanation has been well accepted for organisms with the XY/XX (or WZ/ZZ) scheme of chromosomal sex-determining mechanism, but has been questioned for organisms with the dominant Y (or W) principle of sex determination. However, even in the latter case, Muller's argument that when a tetraploid arises it will usually have to breed with diploids and will thus obliterate itself should also apply. Of course, muller's principle cannot explain the rarity of tetraploids in animals such as insects and reptiles that are capable of producing parthenogenetic polyploids; the main reason here is probably because in animals asexual reproduction is generally less successful than sexual reproduction.] In amphibians and fish the chromosomal determiners of the opposite sexes are still in a rather initial state of differentiation, and the X and the Y or the Z and the W can substitute for each other. In these animals genome duplication would not result in sexual imbalance, and many tetraploid species have been found. Furthermore, Bogart and Wasserman (1972) propose that in amphibians tetraploids can evolve through an intermediate triploid state. Such a situation cannot occur in animals with a well-developed chromosomal sex-determining mechanism.

Complete and Partial Chromosomal Duplication

The duplication of a chromosome or part of a chromosome is likely to cause a severe imbalance in gene product, and its chance of incorporation into the population is small. For example, in *Drosophila* the viability of terminal triploids declines with the amount of the triplicated material, becoming generally lethal when more than one-half of an autosomal arm is present in three doses. In humans, trisomies of larger chromosomes are lethal conditions, and even those of smaller ones cause sterility. Well known examples include such frequent abnormalities as Down's syndrome (trisomy 21), and trisomy of chromosome 18. Similar deleterious manifestations are associated with duplications of parts of chromosomes. Thus, polysomy and partial polysomy have apparently contributed little to evolution, except in some plants such as *Datura stramonium*.

Transposition

As described already, transposition can be duplicative or conservative. In the former case, the original copy is retained at the original site while a new copy is inserted elsewhere in the genome. Thus, duplicative transposition increases the copy number of the transposable element. In conservative transposition, the sequence is excised and reinserted elsewhere, and so there is no increase in copy number. However, a circular intermediate may form from the excised sequence and replicate autonomously. In this case, the number of copies of the element can increase. Not also that transcribed sequence may undergo reverse transcription and reinsertion, i.e., retroposition. If this occurs in the germline, there will be an increase in the copy number of the sequence. As noted, the Alu sequences and many other processed pseudogenes have been produced by retroposition.

It has been suggested that all the middle-repetitive DNA fraction in eukaryotes has originated in transposable elements. Most of the elements are however, no longer mobile, because their ability to transpose has been destroyed by mutations or the insertion of other elements. A large part of the heterochromatin in *Drosophila*, for instance, many in fact be a graveyard of such dead elements.

Comparison of dispersed repeats in related species suggests that transposition and retroposition can rapidly increase the copy number of transposable elements. For example, the genome fraction of dispersed repeats is only about 3% in *Drosophila simulans* but about 21% (24 million bp) in *D. melanogaster*. This large difference in the number of dispersed repeats between the to sibling species is likely to have been mainly due to a rapid increase by transposition rather than due to loss of elements, because excision or loss of transposable elements usually does not occur rapidly. It has also been noted that the number of Alu sequences has increased rapidly by retroposition during the evolution of higher primates.

Replication Slippage and Unequal Crossing-Over

As noted in the preceding section, unequal sister-chromatid exchange seems to be the primary mechanism of mutation at minisatellite loci, while replication slippage seems to be the primary mechanism of mutation at microsatellite and STR loci. Interchromosomal unequal crossing-over is commonly believed to be an important mechanism for the creation of multigene families; however, there is evidence that unequal sister-chromatid exchange is also an effective means for recombination between genes in a family.

There is evidence that the copy number at a VNTR locus can increase rapidly. For example, in humans an allele at the MS32 minisatellite locus may contain more than 600 repeats, whereas the locus is monomorphic with 3-4 repeats in great apes, Old World monkeys, and New World monkeys. The latter organization presumably represents the relatively stable, ancestral precursor state and the high number in some alleles in humans represents a large increase in recent time.

DNA Amplification

It has been noted that replication slippage and unequal crossing-over tend to remove tandem array more often than to increase their size and copy number, and so there mechanisms cannot explain the existence of all the localized repeated DNA. In order to explain the existence of localized, highly repetitive sequences such as satellite DNA, DNA amplification has been proposed. *DNA amplification* refers to any event that increases the number of copies of a gene or a DNA sequence for above the level characteristic for an organism. In particular, it refers to events that occur within the lifespan of an organism and cause a sudden increase in the copy number of a DNA sequence.

One of the most powerful methods of amplification is the *rolling-circle model* of DNA replication. This type of replication is used in the amplification of rRNA genes in amphibian oocytes. Amplification in this case involves the formation of an extrachromosomal circular copy of a DNA sequence, which can then produce many additional extrachromosomal units containing tandem repeats of the original sequence. If such units become integrated back into the genome, there will be an addition to the genome consisting of identical repeated sequences.

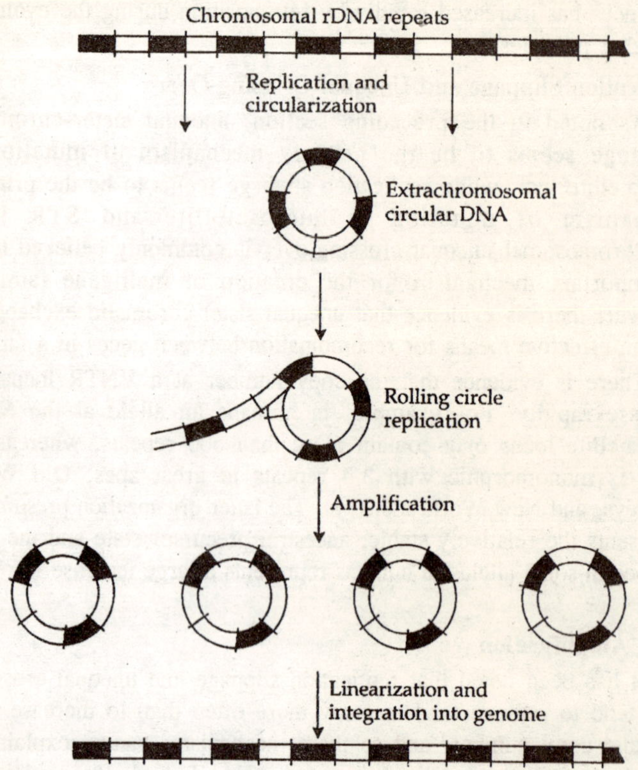

Fig. 2.5. The rolling-circle model of gene amplification in amphibian oocytes.

Maintenance of Nongenic DNA

Solving the C-value paradox and accounting for the structure of the eukaryotic genome require finding evolutionary mechanisms for the long-term maintenance of vast quantities of nongenic, seemingly superfluous DNA. This, in turn, is intimately linked with the question of what function this DNA might have, if any. Numerous attempts have been made to provide evolutionary explanations for this phenomenon. The following are four such hypotheses:

1. The nongenic DNA performs essential functions, such as global regulation of gene expression. According to this hypothesis, the excess of DNA is only apparent, and the DNA is wholly functional. Consequently, deletions or removal of such DNA will have a deleterious effect on fitness.
2. The nongenic DNA is useless "junk", carried passively by the chromosome merely because of its physical linkage to functional

genes. According to this view, the excess DNA does not affect the fitness of the organism and thus will be carried indefinitely from generation to generation.

3. The nongenic DNA is a functionless "parasite" or "selfish DNA" that accumulates and is actively maintained by intragenomic selection.
4. DNA has a structural or nucleoskeletal function, i.e., a function related to the determination of nuclear volume but unrelated to the task of carrying genetic information.

There is very little evidence for the first hypothesis. In fact, most indications are the most nongenic DNA is indeed devoid of function and can be deleted without discernable phenotypic effects. It also seems that excess DNA in eukaryotes does not greatly tax the metabolic system and that the cost in energy and nutrients of maintaining and replicating large amount of nongenic DNA is not excessive. It is thus possible that most nongenic DNA is indeed junk or selfish DNA.

However, there may be some drawbacks in maintaining a large amount of nongenic DNA. First, large genomes have been found to exhibit greater sensitivity to mutagens that small genomes. Second, maintaining and replicating a large amount of DNA may impose a certain burden on the organism. Indeed, the fact that the genome of yeast contains much fewer introns than do those of animals and plants suggests that it is advantageous for yeast to have a streamlined genome. Also, some degree of streamlining in genome size seems to be advantageous for *Drosophila* and other invertebrates, for their intron lengths are, on average, considerably shorter than those in mammals; for example, more than half of the 209 *Drosophila* introns surveyed in one study were shorter than 80 bp, whereas such short introns are rare in mammals.

As mentioned in the preceding section, most middle-repetitive sequences are believed to be transposable elements or their degenerated variants. This class of repetitive sequences would have their origin as selfish DNA. Note that the sequence of a transposable element can become so degenerated by mutation that it is not longer recognizable as a transposable element. It is quite possible that a substantial fraction of the genome of higher eukaryotes is due to such sequences, which would indeed be junk in the sense of the word.

All satellite DNA families appear to be transient in existence, for no satellite family has been found to have persisted for a long evolutionary time. For example no satellite family is shared by all members of the *Drosophila melanogaster* species group, suggesting that every satellite that existed in the common ancestor of this *Drosophila* group has become lost in one or more species of the group. The transient nature of satellite DNA argues against the existence of any significant functional role for satellite DNA. This view is also consistent with the observation that most satellite DNAs are found in chromosomal regions with a low rate of recombination (e.g., heterochromatin). This observation suggests that satellite DNAs are dispensable and that a satellite is likely to persist longer if it is in a region with reduced recombination, for in such a region the chance of being deleted by recombination is lower.

In addition to satellite DNA, families of localized, moderately repetitive DNA are also good candidates for junk DNA. It is possible that most families of localized simple-repetitive sequences are devoid of function and they neither lower nor increase the fitness of their carrier. Consequently, the evolution of such sequences is not affected by natural selection but is determined mainly by gene conversion and unequal crossing-over. These mechanisms will result in two outcomes: (1) sequence homogeneity and (2) wide fluctuations in numbers over evolutionary times. It has also been suggested that the rate of turnover of localized arrays of repetitive sequences is quite high; that is, existing arrays may be removed by unequal crossing-over, while new arrays are continuously created by processes of DNA duplication.

Of course, not all tandemly repeated simple sequences are junk, for there are some families of such sequences that do have a function. As mentioned already, telomeres and centromeres are such examples. Telomeres are short repetitive sequences at the end of eukaryotic chromosomes and provide protection for the end of a chromosome against exonucleotic degradation. Centromeres, which consist of short repetitive sequences similar to those of telomeres, provide specific sites for attachment of the chromosome to segregation machinery during meiosis and mitosis.

Another interesting example of functional tandem repeats is the *Responder* (Rsp) locus in natural populations of *Drosophila melanogaster*. This locus consists of 20-2500 copies of an AT-rich 120-bp-long sequence. In a competition experiment involving a mixed population consisting of flies containing 700 copies of the repeat and flies containing 20 copies, it was observed that the frequency of the repeat and flies containing 20 repeats decreased with time. Therefore, it has been concluded that flies containing 700 copies have a higher fitness than flies containing only 20 copies. At present time, the function of the Rsp locus is not known, but it is apparently not junk DNA since its length affects the fitness of the files.

Most functional repetitive sequences require some sequence specificities and may be classified as "genic" DNA in the broad sense. The skeleton-DNA hypothesis proposes a function for nongenic DNA that requires no sequence specificity but depends only on the amount of DNA. It postulates that DNA acts directly as a nucleoskeleton that is the primary determinant of interphase nuclear volume and that the changes in nuclear volume occurring in different diploid somatic cells of multicellular organisms are brought about primarily by controlling the degree of folding or unfolding of the DNA. Since larger cells require large nuclei, selection for a particular cell volume will secondarily result in selection for a particular genome size. According to this scheme, excess DNA is maintained by selection, but its nucleotide composition can be changed at random. Whether nongenic DNA has the proposed versatility of controlling nuclear volume for different types of cells remains to be tested by experiments.

No single explanation is likel· to solve the C-value paradox. All the above mechanisms, and some additional ones, may contribute to the maintenance of "excess" genomic size, and our task in the future will be to determine the relative contribution of each.

GC Content in Bacteria

Among eubacteria, the mean percentage of guanine and cytosine in genomic sequences, or the *(GC) content* varies from approximately 25% to 75%. Bacterial GC contents are phylogenetically correlated, with closely related bacteria having similar GC contents.

There are essentially two types of hypotheses to explain the variation in GC content in bacteria. The selectionist view regards the GC content as a form of adaptation to environmental conditions. For example, in thermophilic bacteria, which inhabit very hot environments, strong preferential usage of thermally stable amino acids encoded by GC-rich codons (e.g., alanine and arginine) and strong avoidance of thermally unstable amino acids encoded by GC-poor codons (e.g., serine and lysine) have been reported. Therefore, GC content may be a trait that is determined by selection. Another selectionist scenario invokes UV radiation as the selective force. Since T-T dimers are produced by radiation, microorganisms in the upper layers of the soil, which are exposed to sunlight should have a higher GC content than bacteria

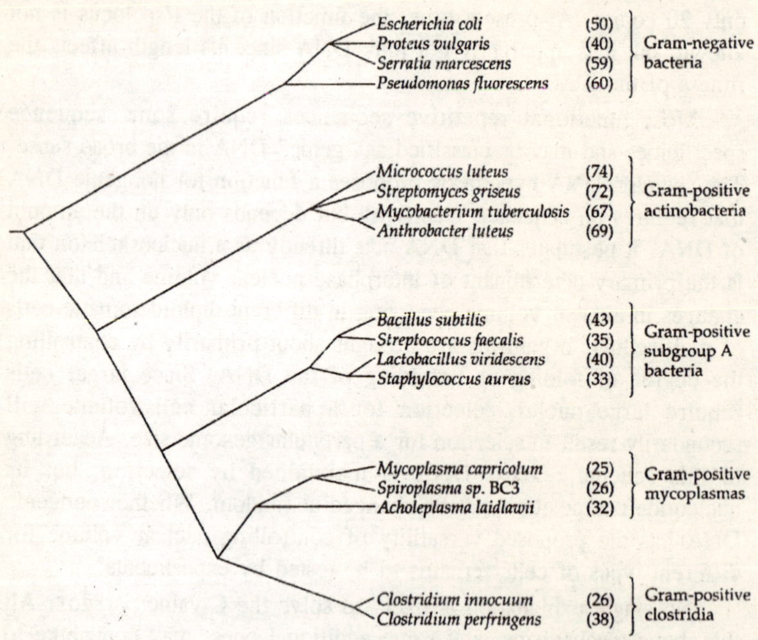

Fig. 2.6. Phylogenetic tree of several eubacteria based on 5S rRNA sequences.

that are not exposed, for example, intestinal bacteria such as *Escherichia coli*.

The mutationalist view invokes biases in the mutation patterns to explain the variation in GC contents. According to this view, the GC content of a given bacterial species is determined by the balance between (1) the rate of substitution from G or C to T or A, denoted as u and (2) the rate of substitution from A or T to G or C, denoted as υ. At equilibrium, the GC content is expected to be

$$P_{GC} = \upsilon / (\upsilon + u) \quad \ldots(1)$$

The ratio u/υ is also called the *GC mutational pressure*. When u/υ is 3.0, the GC content at equilibrium will be 25%. Such is the situation in *Mycoplasma capricolum*. When the ratio is 1, the GC content will be 50%, as in *Escherichia coli*. When it is 0.33, the GC content will be 75%, as in *Micrococcus luteus*.

In addition to the GC mutational pressure, mutational changes are subject to selective constraint. The weaker the constraint is in a particular region, the stronger the effect the GC mutational pressure will be on the GC composition. Fig. 2.7 shows the correlation between the total GC content and the GC content at the three codon positions

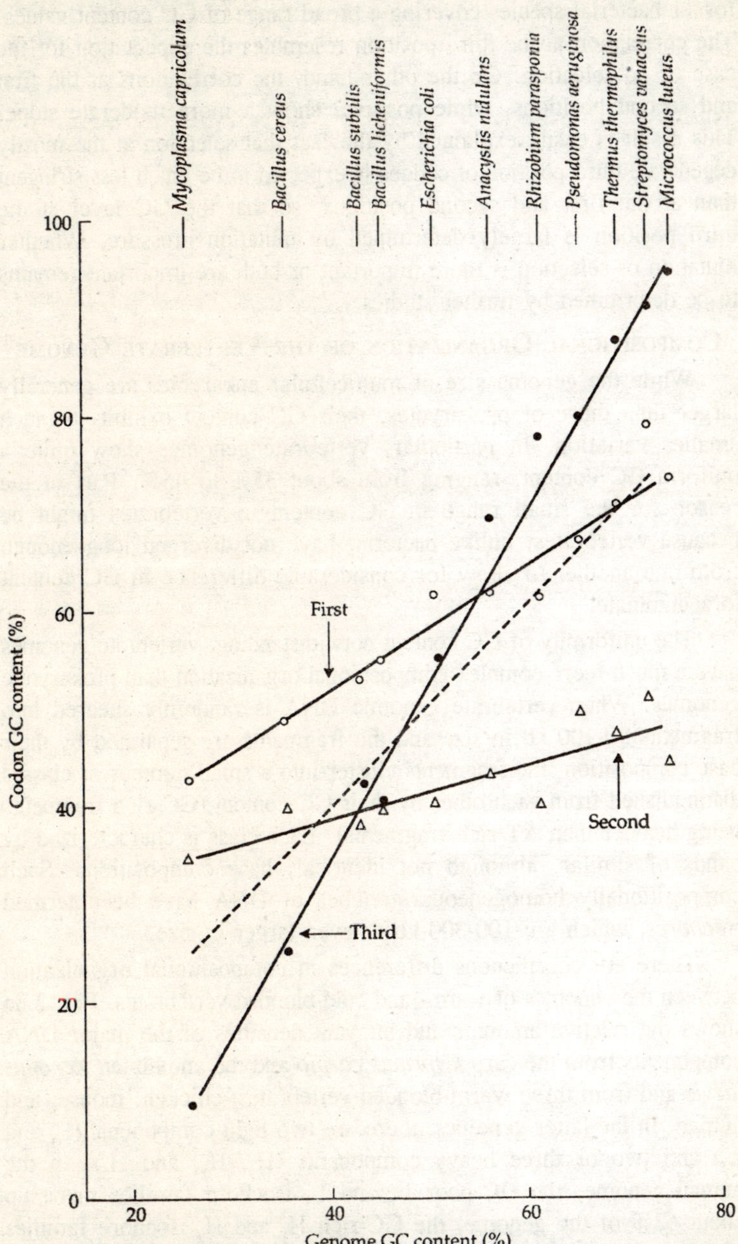

Fig. 2.7. Correlation of the GC content between total genomic DNA and the first (open circles), second (open triangles), and third (solid circles) codon positions.

for 11 bacterial species covering a broad range of GC-content values. The correlation at the third position resembles the expectation for the case of no selection. On the other hand, the correlations at the first and second positions, while positive, show a more moderate slope. This result is easily explained by the fact that selection at the mostly degenerate third position of codons is expected to be much less stringent than at the first and second positions, so that the GC level at the third position is largely determined by mutation pressure. Whether mutation or selection is more important or both are important remains to be determined by further studies.

Compositional Organization of the Vertebrate Genome

While the genome size of multicellular eukaryotes are generally larger than those of prokaryotes, their GC content exhibits a much smaller variation. In particular, vertebrate genomes show quite a uniform GC content, ranging from about 35% to 45%. Part of the reason for the small range in GC content in vertebrates might be because vertebrates, unlike bacteria, have not diverged long enough from one another to allow for considerable difference in GC content to accumulate.

The uniformity of GC content notwithstanding, vertebrate genomes have a much more complex compositional organization than prokaryotic genomes. When vertebrate genomic DNA is randomly sheared into fragments 30-100 kb in size and the fragments are separated by their base composition, the fragments cluster into a small number of classes distinguished from each other by their GC content (GC-rich fragments being heavier than AT-rich fragments). Each class is characterized by bands of similar, although not identical, base compositions. Such compositionally homogeneous stretches of DNA have been termed *isochores*, which are 100-300 kb or even larger in size.

There are conspicuous differences in compositional organization between the genomes of warm- and cold-blooded vertebrates. Fig. 2.8a shows the relative amounts and buoyant densities of the major DNA components from the carp *Cyprinus carpio* and the amphibian *Xenopus laevis* and from three warm-blooded vertebrates: chicken, mouse, and human. In the latter genomes, there are two light components (L_1 and L_2) and two or three heavy components (H_1, H_2, and H_3). In the human genome, the GC-poor L_1 and L_2 isochore families make up about 62% of the genome, the GC-rich H_1 and H_2 isochore families represent about 22% and 9%, and the GC-richest isochore family H_3 represents about 3-4%; the remaining part of the genome is formed by

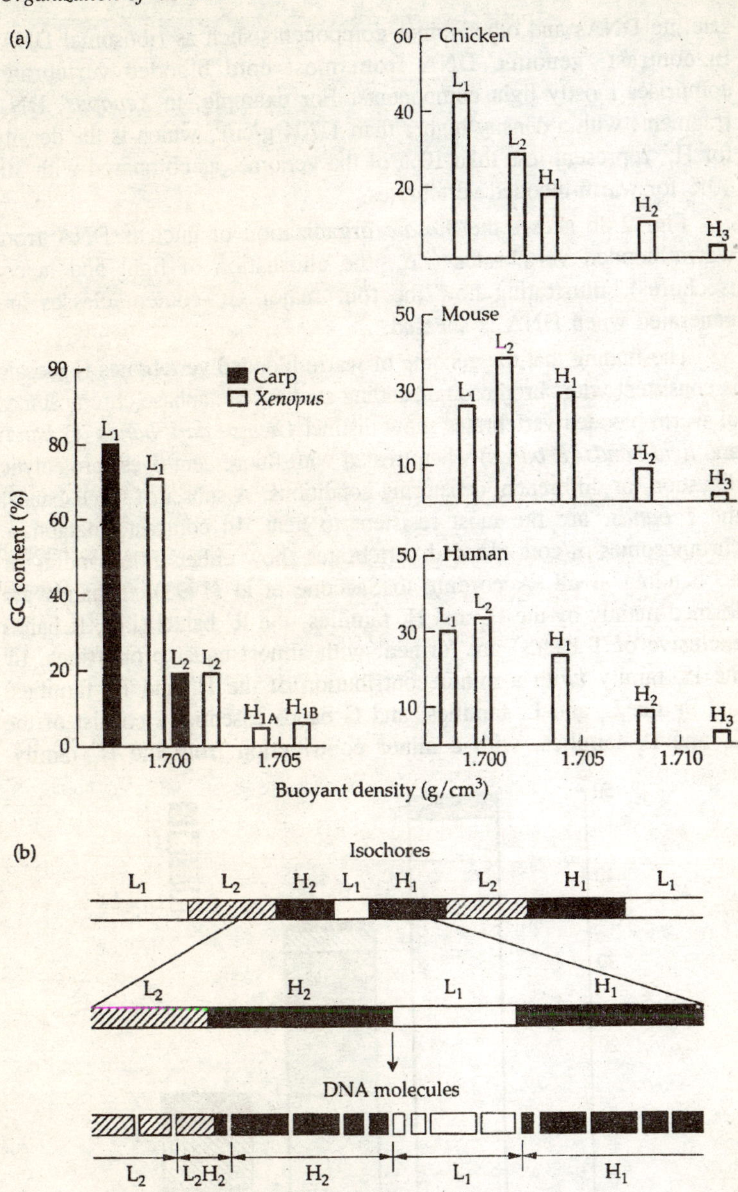

Fig. 2.8. (a) Histograms showing the relative amounts and buoyant densities of the major DNA components from Cyprinus carpio and Xenopus laevis (left panel) and from chickens, mice and humans (right panel). (b) Scheme depicting the mosaic organization of nuclear DNA from warm-blooded vertebrates.

satellite DNAs and other minor components such as ribosomal DNA. In contrast, genomic DNA from most cold blooded vertebrates comprises mostly light components. For example, in *Xenopus*, DNA fragments with a density higher than 1.704 g/cm^3, which is the density for H_1, represent less than 10% of the genome, as compared with 30-40% for warm-blooded vertebrates.

Fig. 2.8b shows the *mosaic* organization of nuclear DNA from warm-blooded vertebrates (i.e., the alternation of light and heavy isochores), illustrating how the four major GC-content classes are generated when DNA is sheared.

The finding that the genome of warm-blooded vertebrates is mosaic is consistent with chromosome-banding studies. Metaphase chromosomes of warm-blooded vertebrates show distinct *Giemsa dark bands* (*G band*) and *light bands* (*R bands*) when treated with fluorescent dyes, proteolytic digestion, or differential denaturing conditions. A subset of the R bands, the *T bands*, are the most resistent to heat. In contrast, metaphase chromosomes of cold-blooded vertebrates show either little banding or no banding at all. According to Saccone et al (1993), T bands are formed mainly by the H_2 and H_3 families, the R' bands (i.e., R bands exclusive of T bands) are formed, with almost equal proportions, by the H_1 family (with a minor contribution of the H_2 and H_3 families) and by the L_1 and L_2 families; and G bands essentially consist of the L_1 and L_2 families, with a minor contribution from the H_1 family.

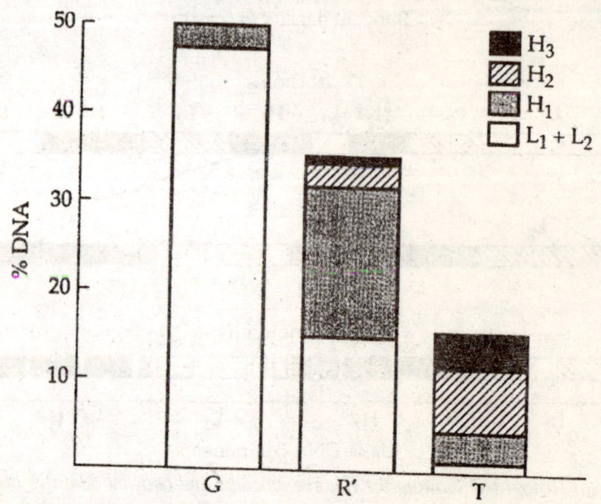

Fig. 2.9. A schematic presentation of the relative amounts of isochore families $L_1 + L_2$, H_1, H_2, and H_3 in the chromosomal bands of the human karyotype.

Studies on the replication timing of genes show that genes localized in GC-rich isochores (T bands) replicate early in the cell cycle, whereas gene localized in GC-poor isochores (G bands) replicate late.

The first few genes localized in compositional DNA fractions suggested that genes are not distributed randomly in the human genome but are concentrated in GC-rich isochores. Subsequent investigations indeed indicated that gene concentration is low in GC-poor isochores, increases with increasing GC in isochores families H_1 and H_2, and reaches a maximum in isochore family H_3, which exhibits up to a 20-fold higher gene concentration compared to GC-poor isochores.

Analysis of DNA sequence data have revealed a positive correlation between the GC levels of genes and the GC level in the large DNA regions in which the genes are embedded. The β and β-like globin genes are low in GC content and are embedded in a low-GC region. The α and α-like globin genes, on the other hand, are GC rich and are embedded in a GC-rich region. The same situation is found in rabbits, goats, and mice. In chickens, both β- and α-globin genes are GC rich, and both are embedded in GC-rich regions. In contrast, the α- and β-globin genes in *Xenopus* are GC-poor, and both are embedded in a GC-poor region.

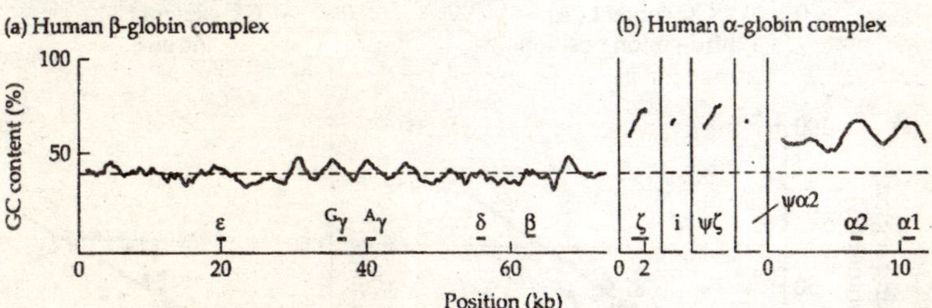

Fig. 2.10. Distribution of GC content along human globin DNA sequences. (a) The β-globin gene cluster; (b) the α-globin gene cluster.

In the vast majority of cases, the GC content in coding regions tends to be higher than that in the flanking regions. We also see that the GC level at the third-codon position is on average higher than that in introns, which in turn is higher than that in the 5' and 3' flanking regions. The GC level in the 5' flanking region tends to be higher than that in the 3' flanking region, probably because the promoter and its surrounding regions tend to be GC rich.

Origins of Isochores

The origin of GC-rich isochores is mysterious and controversial. Note that what is at issue is the general tendency of long DNA

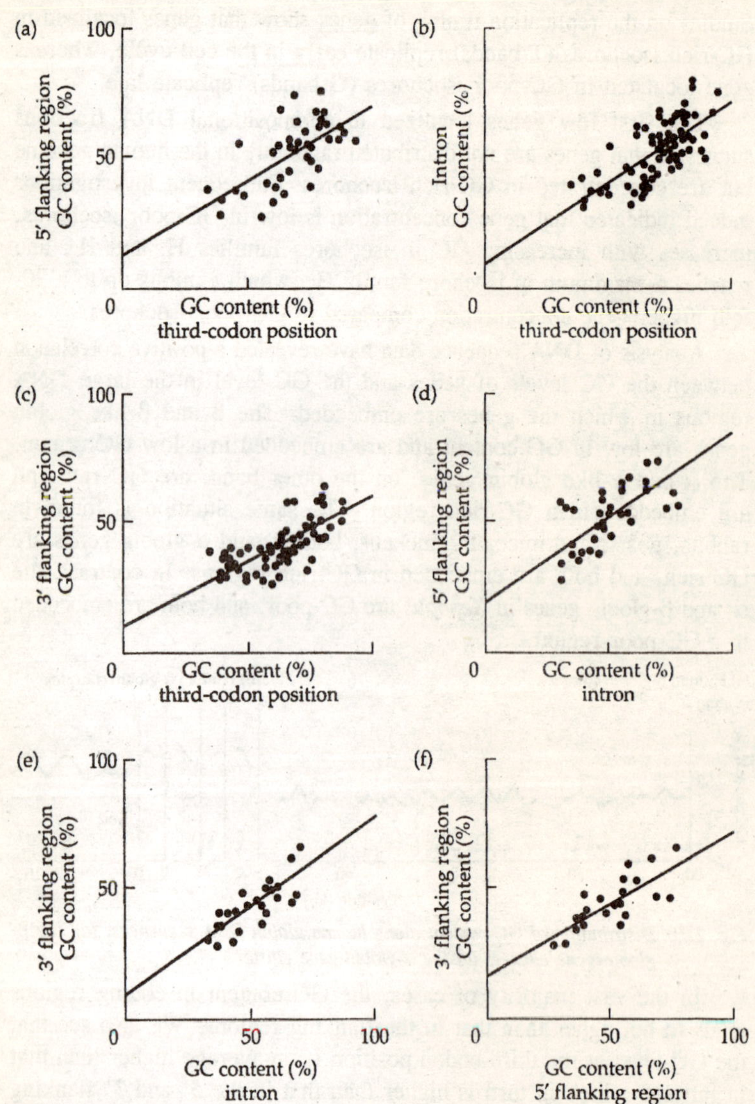

Fig. 2.11. Relationships among the percentage of GC in the various regions of the gene. (a) The third-codon position and the 5' flanking region; (b) The third-codon position and introns; (c) The third-codon position and the 3' flanking region; (d) Introns and the 5' flanking region; (e) Intron and the 3' flanking region; and (f) 5' and 3' flanking regions.

segments (100 kb or longer) to be homogeneous in GC content, not the localized variation in GC content, such as that observed among the various regions of a gene. Bernardi et al. (1985, 1988) proposed that isochores arose because of a functional (i.e., selective) advantage. Their main argument is that in warm-blooded organisms an increase in GC content can protect DNA, RNA, and proteins from degradation by heat because G-C bonds are stronger than A-T bonds. We call this view the selectionist hypothesis.

The opposing view is that GC-rich isochores arose because of mutational pressure. In particular, the hypothesis proposed by Wolfe et al. (1989a) stipulates that isochores arose from mutational biases due to compositional changes in the precursor nucleotide pool during the replication of germline DNA. The GC-rich isochores are carried on DNA regions that replicate early in the germline cell cycle during which the precursor pool has a high GC content and, thus, a propensity to mutate to G or C. The AT-rich isochores, on the other hand, are replicated late in the cell cycle, when the precursor pool has a high AT content and a propensity to mutate to AT. We call this the mutationalist hypothesis. This hypothesis is based on the observations that the composition of the nucleotide precursor pool changes during the cell cycle and that such changes can in fact lead to altered base ratios in the newly synthesized DNA. Note that the replication of the mammalian genome is quite a lengthy process, taking eight hours or more recently showed that, under suitable conditions, mutation pressure can indeed lead to isochore formation.

An argument for the selectionist hypothesis is that increases in GC content at the first- and second-codon positions may confer thermal stability to proteins, while increases in GC content in introns, third-codon positions, and untranslated regions can both increase the thermal stability of primary mRNA transcripts and stabilize chromosomal structures, possibly through effects on DNA-proteins interactions. Indeed, in thermophilic bacteria, strong preferential usage of GC- rich codons has been reported. However, the body temperature of worm-blooded vertebrates is much lower than the temperatures experienced by thermophilic bacteria, so temperature may not be a very important factor in the evolution of proteins and DNA sequences in vertebrates.

One difficulty with the selectionist hypothesis is the fact that a substantial proportion of the mammalian and avian genes are low in GC content. This hypothesis also cannot explain why some duplicate genes have opposite GC contents. For example, in mammals, the β-globin cluster is low in GC, while the α-globin cluster is GC-rich, although the two types of globin genes are expressed in the same cells and serve the same function. Another example is the insulin-like growth factors I and II (IGF-I and IGF-II) genes; in humans IGF-II is closely liked to the insulin gene and located on chromosome 11 whereas IGF-I is located on chromosome 12. In cold-blooded vertebrates, insulin and the IGFs are similar in GC content. In contrast, insulin and IGF-II demonstrate dramatic increases in GC content in mammals, though no such trend occurs in IGF-I; for example, the GC content in introns is only 40-43% in human IGF-I but higher than 60% in the IGF-II and insulin genes. Similarly, some immunoglobulin genes are located in GC-rich regions while others are located in AT-rich regions. Bernardi et al.'s (1985, 1988) explanation is that isochores represent the unit of selection and the α-globin cluster in mammals has been translocated to a high-GC isochore, whereas the β cluster remained in a low-GC isochore. According to this argument, the α and β clusters in chickens should have both been translocated to GC-rich isochores because both of them have a high-GC content. This argument, however, raises the issue of the functional advantage of a GC-rich isochore. If the advantage does not come from the genes it contains, where does it come from?

The mutationalist hypothesis can explain the large difference in GC content between the α- and β-globin clusters in the mammalian genome by assuming that they are located in an early- and late-replicating region (replicon), respectively. However, this hypothesis also has difficulties. First, it is difficult to explain why GC-rich isochores rarely occur in cold-blooded vertebrates. Second, it is not clear why in both the avian and mammalian lineages the increase in GC content occurred preferentially in regions with higher gene concentrations. Of course, it is possible that the majority of such regions happened to be on early replicons. Third, constitutive heterochromatin, such as satellite DNAs, which are mostly GC-rich, and facultative heterochromatin, such as the inactive X chromosome, replicate at the end of the cell cycle, and in these cases, no connection between changes in nucleotide pools and DNA composition in observed.

In conclusion, the presently available data seem to be insufficient to distinguish between the two hypotheses. It is also possible that both mutation pressure and natural selection have played roles in shaping the compositional organization of the genome of warm-blooded vertebrates. One line of future research for resolving the controversy is to examine the degree of discontinuity in GC content at isochore borders. The isochore border between an L (GC-poor) and an H_2 (GC-rich) isochore in the MHC region was recently sequenced and found to have a sharp compositional discontinuity. Limited available data suggest that this represents the general case. Such discontinuities can be explained by the mutationist hypothesis by assuming that a discontinuity occurs because the isochores on the two sides of the border are located on replicons that are replicated at very different times during germline DNA replication. Incidentally, it is interesting that the boundary sequence in the MHC region is highly similar to the sequence at the pseudoautosomal boundary on the human X and Y chromosomes.

Origins of Introns

The discovery of intervening sequences (introns) in eukaryotic genes was a great surprise and soon raised questions about their origin and significance in evolution. Gilbert (1978) suggested that this gene organization could speed up evolution by providing mechanisms for the generation of novel proteins from old ones (i.e., exon shuffling), Doolittle (1978) and Darnell (1978) speculated that the "genes-in-pieces" structure is a primitive form that was present in the genome of the progenote, and Blake (1978) suggested that 'genes-in-pieces" might imply "proteins-in-pieces," that is, exons originally corresponded to structural units of proteins. This view has been known as the introns-early hypothesis. Later, Gilbert (1987) modified it to the exon theory of genes, proposing that exons are the descendants of ancient minigenes and introns are the descendants of spacers between them. The opposing view, i.e., the introns-late view, assumes that early genes had no introns and that the addition of introns and splicing mechanism occurred after the emergence of the eukaryotic cell or the emergence of mitochondria. The debate between the two schools has been lively and fascinating—it has taken several turns in the past two decades, with the weight of evidence favouring sometimes one view and sometimes the other.

Before discussing the debate, we define *cenancestor* and *progenote*. *Cenancestor* refers to "the most recent common ancestor to all organisms that are alive today". In the view that the eukaryotes are derived from archaebacteria, the cenancestor refers to the common ancestor of the eubacteria and archaebacteria. *Progenote* is Woese and Fox's (1977b) descriptor for early biological entities that had only "a rudimentary, imprecise linkage between its [their] genotype and phenotype" and were much simpler than contemporary prokaryotic cells. Originally, they thought that the eukaryotic lineage (cytoplasm) was very ancient and that it was at the progenote stage, not the prokaryote stage, that the line of descent leading to the eukaryotic cytoplasm diverged from the bacterial lines of descent. In this sense, the progenote would be the cenancestor. Woese and Fox's view is not supported by current data, and the term progenote is now commonly used to refer to the most primitive organism (probably with an RNA genome) from which all living cells have descended; it is probably considerably more ancient than the cenancestor.

The debate is on the origin of the classical nuclear *spliceosomal introns*, often called *nuclear introns*. They are different from group I and group II introns, which have self-splicing activities. In the introns-late view, spliceosomal introns were derived either form transposable elements or from group II introns. In the latter proposal, some group II introns from mitochondria or chloroplasts invaded the nuclear genes of the eukaryotic cell by retroposition and evolved into the nuclear introns we see today. This may be called the mitochondrial (or

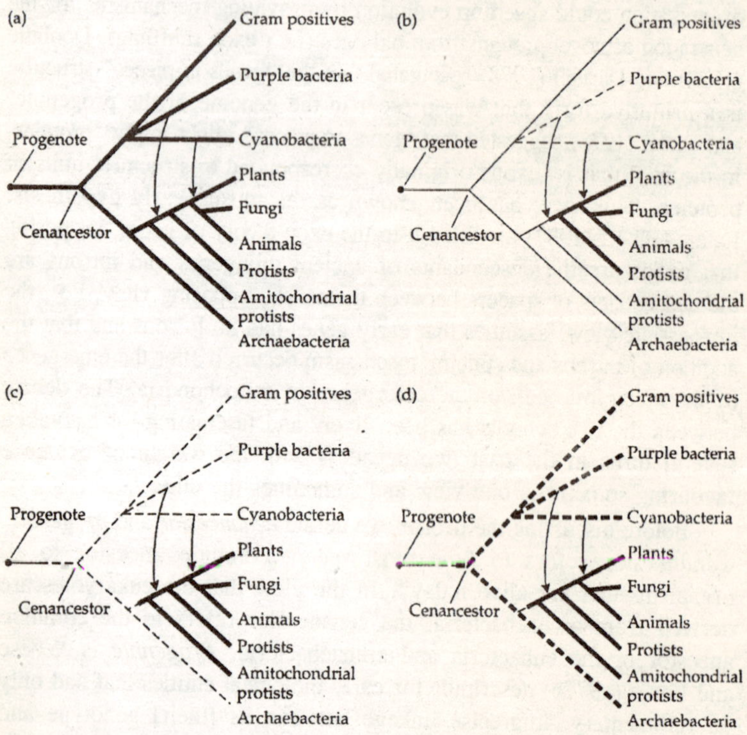

Fig. 2.12. Schematic representations of four possible senarios for the evolution of introns. (a) The progenote origin of introns; (b) The mitochondrial origin of introns; (c) The nuclear origin; (d) This prokaryotic origin model has the degeneration of group II introns in a prokaryote before the divergence of eukaryotes.

organelle) origin of introns. On the other hand, the introns-early view states that spliceosomal introns are descendants of self-splicing introns (ribozymes) that were present in the progenote. After the divergence of the eubacterial, archaebacterial, and eukaryotic lineages, most or all of these ancient introns were lost in the Eubacteria and Archaebacteria, whereas in the eukaryotes these introns evolved to the spliceosomal introns. Note that the exon theory of genes and the introns-early hypothesis have commonly been mistaken as equivalent, but actually they are not, because introns might be present in the progenote but they were not derived from spacers as Gilbert (1987) proposed. Note also that in addition to these two scenarios, there are other possible alternatives. One of them is that group II introns were present in all or most prokaryotes before the emergence of eukaryotes and were inherited by the nucleus, and degenerate into spliceosomal introns in eukaryotes (the nuclear origin hypothesis). Another scenario is that some group II introns evolved into spliceosomal introns in a prokaryote before the divergence of the eukaryotes.

The presence or absence of introns in the progenote or early primitive organisms has profound implications for the evolution of early genes. In the introns-early view, the primordial introns played an important role in the early assembly of proteins and accelerated the early evolution of proteins by exon-shuffling. In contrast, in the introns-late view, there was no spliceosomal intron in the genome of he progenote and so exon shuffling played no role in the assembly and evolution of early genes; self-splicing introns might have emerged early in evolution, but such introns have structural requirements that make them less suitable for exon-shuffling than we are spliceosomal introns.

Relationships between Exons and Units of Protein Structure

One line of evidence that has been used to support the introns-early view is the correlation between modules and exons. For example, Go (1981) identified four modules, F1-F4, in the α and β globins, showed that F1, F2+F3, and F4 correspond to exons 1, 2, and 3 in vertebrate α- and β-globin genes, and predicted the existence of an intron at a position between F2 and F3 in an ancient globin gene. The finding of a third intron exactly at the predicted region in the gene of leghemoglobin from soybean was taken as strong evidence for the exon theory of genes. A similar prediction of the existence of an intron at a particular region in the triose phosphate isomerase (TPI) gene was made and found to be true. However, as more sequences of a gene from diverse organisms become available, the number of introns increases, and the correlation between exons and protein structure units tends to decrease or even disappear. For instance, 12 rather than 3 introns have later been found at different locations of globin-related genes, and when these 12 introns positions were analyzed, no correlation between exons and units of protein structure was found. The same conclusion was reached for alcohol dehydrogenase, pyruvate kinase, and TPI. In response, Long et al. (1995a) argued that there are two problems with this type of analysis. The first, and major, one is the definition of ancestral introns, because today's intron/exon structures are the final products of intron loss, intron sliding, and possibly intron gain. The second problem is that sequence alignment maybe problematic—in comparing divergent genes. there can be regions that lack enough sequence similarity to make a reliable alignment. Thus, what appear to be different intron positions in such regions may only reflect alignment artifacts. Obviously, the two schools hold very different views on this issue.

The introns-early/introns-late controversy is unlikely to be resolved by studying the relationships between exons and structural units of proteins for several reasons. First, it involves the issue of intron gain and loss. Obviously, a change in intron number either by loss or gain can reduce the correlation between exons and structural units. Intron gain is a controversial but important issue and will be discussed again later. Second, it involves the issue of *intron-sliding*, a shift in the position of an intron in the gene, a situation that may arise from mutations that create a new splice junction and block the old one. If sliding had occurred, the resultant intron might be mistaken for a new intron and the correlation between exons and structural units would be reduced; however, Stoltzfus et al. (1994, 1995) test may still be valid, if the sliding is small. Intron-sliding has been invoked to infer homology of introns located at slightly different positions in different organisms, but this approach has been challenged by Logsdon et al. (1995). A good candidate for intron-sliding has been proposed for two duplicated genes in *Volvox*. Third, an ancient structural unit in a gene may degenerate if it becomes no longer useful to the gene. In this case, the correlation between exons and structural units is reduced. Fourth, it is difficult to know whether a structural unit such as an α helix or a β sheet in a gene is ancient or derived. Finally, although absence of a correlation between exons and protein structural units may be taken as evidence against the exon theory of genes, it can not refute the introns-early view, because, as noted above, introns might be present in the progenote but had nothing or little to do with protein structure. Note also that this type of study cannot distinguish among the three scenarios.

Phylogenetic Distribution of Introns

Another line of evidence used to support the introns-early vie is the correspondence of intron positions between plant and animal genes. For example, four TPI introns were shared by the plants and animals studied. However, when more diverse organisms including protists were studied, the phylogenetic distribution of introns became difficult to interpret. For example, only a few of the dozens of introns in genes for actin and α and β tubulins were found to be widely shared by plants, animals, fungi, and late-arising protists, and it was argued that such distribution are difficult to explain under the introns-early view, whereas each of them could be explained by a single recent gain without subsequent loss. Later, several genes, e.g., glyceraldehyde-3-phosphate dehydrogenase (GAPDH) and RNA polymerase II large subunit, were found to lack introns in some, if not all protists, but contain many introns in animals, plants, and fungi. A summary of extensive data which reveals a complete absence of nuclear introns from all examined genes of the earliest protistan lineages. Palmer and Logsdon (1991) took this as strong evidence that nuclear introns are relatively recently derived features of eukaryotes. Fig. 2.13 also shows that introns have increased markedly in number of plant, animal, and fungal lineages relative to those (late) protistan lineages that do contain introns. To explain the observed intron distribution, the introns-early view would have to postulate the parallel loss of tens of thousands of introns from many different protistan lineages, including complete introns extinction from the several earliest lineages.

Although the observed phylogenetic distribution of introns poses a serious difficulty for the introns-early hypothesis, it cannot refute the hypothesis for two reasons. First, if intron loss is selectively advantageous in prokaryotes and protists because of selection for an increased rate of genome replication (i.e., selection for genome streamlining), then most introns may indeed become lost in these organisms, given their long history of evolution. There is strong evidence that introns loss can indeed occur, for example, the apolipoprotein A-IV gene has lost the first of its three introns and the mouse insulin I gene has lost the second of its two introns. However, just how easily an intron can become lost is a subject of dispute. Second, intron gain may be selectively disadvantageous in lower organisms (for the same

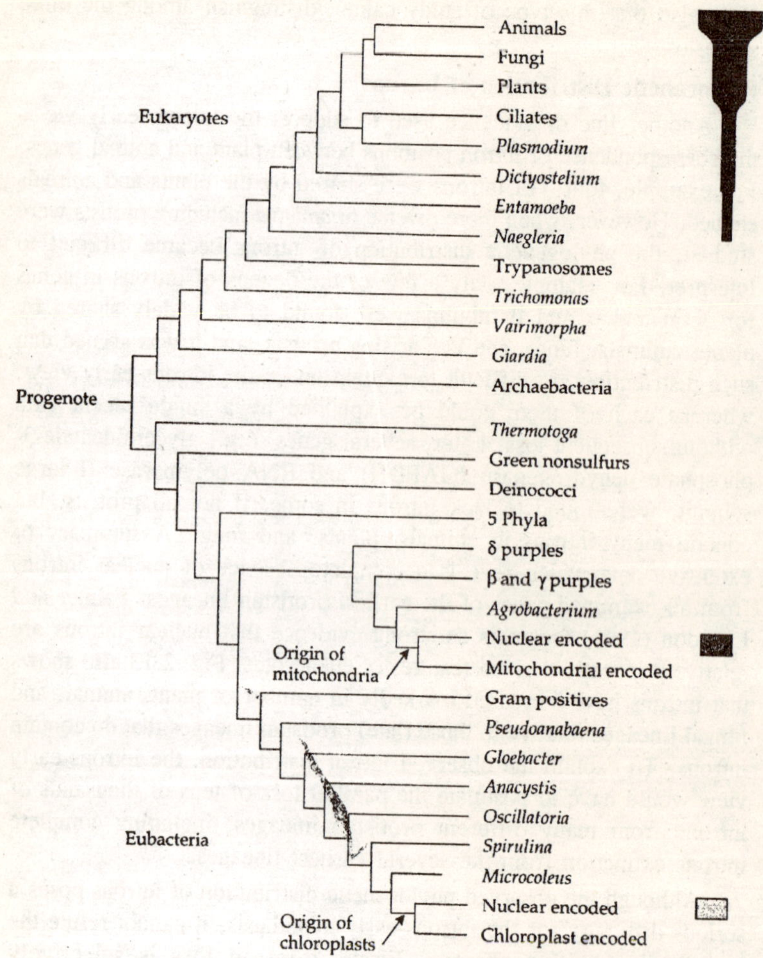

Fig. 2.13. Sporadic distribution of nuclear introns on the tree of life.

reason as above), but not so, or less so, in higher organisms. Note also that intron gain can spread more easily in sexual organisms than in asexual organisms, for one of two reasons. If introns are junk or selfish DNA, they may spread in a population even if deleterious, but only if outbreeding events are regular enough. Alternatively, the existence of introns may reduce the chance of mispairing and thus the rate of recombination between paralogous sequences located at nonhomologous chromosomal sites. Of course, it is difficult to know whether the above factors together can produce the phylogenetic distribution. There is, however, another possible factor that may increase the number of introns in higher organisms.

Another issue concerns the interpretation of the existence of introns at the same positions in nuclear genes and homologous mitochondria- or chloroplast-derived genes now present in the nuclear genome. Some such examples are: (1) five introns at identical positions between nuclear and chloroplast GAPDH, (2) two identical introns between cytosolic and mitochondrial malate dehydrogenase, and (3) five identical introns between cytosolic and mitochondrial aspartate aminotransferase. In the introns-early view, these examples represent ancient introns that existed before the divergence between eukaryotes and prokaryotes and are not due to coincidence resulting from nonrandom insertions into so-called protosplice sites, as suggested by the introns-late school. However, Palmer and Logsdon (1991) have challenged the claim that all of these examples represent introns at identical positions and have argued that because chloroplasts (with a similar argument for mitochondria) emerged so late in evolution (~1 billion years ago), the introns-early hypothesis would force one to postulate retention of these introns for two-thirds of eubacterial evolution (before the derivation of chloroplasts from cyanobacteria), as well as their parallel and independent loss in many separate eubacterial lineages. They therefore favoured the interpretation of parallel insertion of introns at the same or similar positions.

Long et al. (1995b) have recently studied the distribution of intron phases in ancient conserved regions (ACRs), which are complete genes or portions of genes that are shared by both eukaryotes and prokaryotes. These regions have no introns in prokaryotes but may have introns in the eukaryotic counterparts. On such introns-late models, these introns must have been inserted. For such models, they could not participate in exon shuffling because the ACR regions are colinear with the ancestral molecules, which on the introns-late hypothesis came into existence before there were introns. Long et al. examined these ACR introns for phase bias and for phase correlation in the same genes. They found that about 55% of the introns studied were of phase zero and that intron phases in the same genes are correlated; in contrast, under the assumption of random insertion, the expected frequency of each intron phase should be 1/3 and there should be no correlation between intron phases in a gene. They found an excess of symmetric exons (exons flanked by the same phase introns) or symmetric sets of exons, significant at the 1% level. They interpreted this excess as evidence of exon shuffling (duplication), arguing that the shuffling would have had to have occurred in the progenote or early organisms, and hence these introns would have existed in the progenote. They therefore took their finding as evidence for the introns-early hypothesis. Note, however, that the putative shuffling actually only had to have occurred before the cenancestor, not necessarily in the progenote. Therefore, Long et al.'s observation is also compatible with the prokaryotic origin of introns.

Mechanisms of Intron Insertion

In the current version of the introns-late hypothesis, nuclear introns are originally derived from group II introns. Recent demonstrations of reverse self-splicing of group II introns *in vitro* suggest a ready mechanism for the insertion of group II introns into nuclear genomes. An intron that was spliced out is self-spliced into an RNA. That intron containing RNA molecule is then later copied by reverse transcriptase into cDNA, and the intron is inserted by recombination into the nuclear gene; this process is similar to the creation of processed pseudogenes or retrosequences. The intron thus created would be a "perfect intron" because it could be precisely excised from the pre-mRNA by self-splicing. Palmer and Logsdon (1991) suggest that nuclear genes are continually invaded by group II introns by this reverse self-splicing mechanism. Rapid degeneration of a group II intron into a spliceosome-dependent form could then occur, perhaps stemming from a single nucleotide substitution that would establish a nuclear-intron consensus splice site. One question is, Where do these group II introns come from? Nothing that many chloroplast and mitochondrial genomes do contain group II introns and that organelle sequences are frequently transferred into nuclear genomes, Palmer and Logsdon (1991) suggest that the organelles have been seeding the nucleus with group II introns. The failure to finding functional group II introns in nuclear genomes is assumed to be due to their relatively rapid degeneration into standard nuclear introns.

However, with respect to more recent intron gains it is not clear that organelles have been seeding the nucleus with enough introns, especially in view of the fact that group II introns appear to be absent from animal mitochondrial genomes and that animals do not have chloroplasts. A more plausible process of intron gain may be the reverse splicing of existing introns, as has recently been proposed for the addition of spliceosomal introns to the U6 small nuclear RNA genes in yeasts, one in *S. pombe*, and five in different species of *Rhodotorula*. An intron thus created is already a perfect spliceosomal intron: it requires no further evolution, and there is no need to find an intermediate form as in the scheme proposed by Palmer and Logsdon (1991). Long et al. (1995a) suggest that this is a general mechanism for the movement or insertion of spliceosomal introns into RNA. They also note from the yeast examples of insertion that there is no evidence of protosplice sites as proposed by Dibb and Newman (1989). Another possible source of recent intron gain is form transposable elements—there is evidence that some transposable elements in maize can be spliced out like an intron.

Note that the preceding process may be able to explain why higher organisms have numerous introns, because the more introns an organism has, the higher the chance for this process to occur. If addition of introns creates no or little selective disadvantage to their carriers, as is commonly thought for higher organisms, the added introns can be retained and spread to the species. Of course, although this may explain why there are far more introns in higher organisms than in lower ones, it does not explain the origin of introns.

In conclusion, although the introns-early view was once the prevailing view and has eve·. been cited as an established theory in textbooks on molecular and cell biology, it is facing serious challenges. In fact, current evidence seems to be more in favour of the introns-late hypothesis rather than the introns-early hypothesis. However, the controversy is still unsettled and may continue for a long time.

3

Reconstruction of Ancient Genes

Proteins form extinct organisms can be studied by the analysis of DNA recovered form preserved organic specimens. This approach, however, requires biological material that has been completely protected from the oxidative effects of oxygen and water and significant doubts may remain as to the authenticity of the DNA recovered. Alternatively, an attempt can be made to reconstruct the amino acid sequence of an ancient protein from the sequences of its extant descendants. This methodology relies upon the principle of maximum parsimony (i.e. the assumption that extant proteins have evolved from that of an extinct ancestor by the smallest number of mutational changes). DNA sequence data can also be treated in the same way, one example being the investigation of ribosomal DNA phylogeny.

Jermann et al., (1995) examined the evolutionary history of artiodactyl RNase A, a pancreatic digestive enzyme, derived the ancestral enzymes phylogenetically and reconstructed them by site-directed mutagenesis. The kinetic properties of the reconstructed enzymes were found to be similar to those of extant RNase but less stable to thermal denaturation, more susceptible to proteolysis and five-fold more active toward double-stranded RNA. A Gly38→Asp substitution, which occurred ~40 Myrs ago around the time when foregut rumination evolved, was found to be associated with the reduction in activity toward double-stranded RNA.

A similar analysis has been attempted with the chymases (mast cell proteases) which hydrolyse angiotensin I to generate angiotensin

II, a potent vasoconstrictor hormone. Primate α-chymases are highly specific and only hydrolyse the Phe8-His9 peptide bond. By contrast, rat β-chymase is less specific and further hydrolyses angiotensin I by cleavage of the Tyr4-Ile5 bond. Chandrasekharan et al. (1996) determined the phylogeny of four mammalian α-chymases and six mammalian β-chymases and reconstructed the putative ancestral enzyme by chemical synthesis. This enzyme proved capable of cleaving angiotensin I at the Phe8-His9 bond but not at the Tyr4-Ile5 bond. Thus the relatively narrow specificity of the primate α-chymase represents the ancestral state of the enzyme while the broad specificity of the rat β-chymase must have been secondarily derived.

Molecular Reconstruction and Homology Modelling of the Catalytic Domain of the Common Ancestor of the Hemostatic Vitamin K-dependent Serine Proteases

Adopting the maximum parsimony principle and employing novel cDNA-based strategy, Krawczak et al. (1996) reconstructed the catalytic domains of the early mammalian ancestors of the vitamin K-dependent factors and then went on to reconstruct the putative common ancestor of all five proteins from an earlier stage of vertebrate evolution. A tertiary model of the ancestral vitamin K-dependent serine protease was built that was both energetically satisfactory and possessed a credible fold, and its topological and biophysical properties were examined Wacey et al. (1997) then traced the evolution of specific structural features in protein C from the common ancestor of the vitamin K-dependent serine proteases toward extant human protein C. These studies will not be described in some detail since they serve to illustrate the potential value of homology modelling in evolutionary studies.

Evolution of the Vitamin K-dependent Coagulation Factors

The vitamin K-dependent serine proteases of coagulation (factors VII, IX, and X, prothrombin and protein C) exhibit substantial sequence and structural homology. Factors VII, IX, X and protein C all contain an N-terminal domain of glutamic acid (Gla) residues, two epidermal growth factor-like (EGF) domains and a catalytic domain; prothrombin differs slightly in that it possesses two kringle domains instead of the EGF domains. With the exception of prothrombin, the genes encoding the vitamin K-dependent coagulation factors have a similar exon/intron organization suggesting that they have arisen from a relatively recent common ancestor through a process of gene duplication and divergence. The organization of these gens also reflects the functional modular

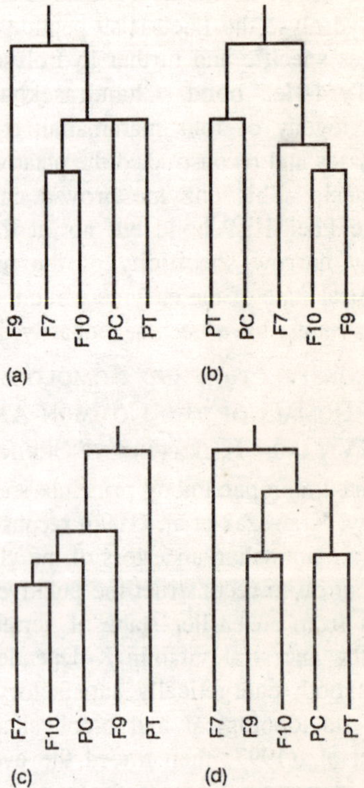

Fig. 3.1. Proposed phylogenies for the vitamin K-dependent factors of coagulation.

assembly of the respective proteins and is thought to have emerged through exon shuffling. Evolution of these genes has proceeded by repeated insertions, duplications, exchanges and deletions of modules. The presence/absence of modules such as calcium-binding Gla domains, the EGF-like domains and kringles was used by Patthy (1990) to reconstruct the evolutionary past of the genes. More recently, it has been realised, however, that each module or domain has its own distinctive evolutionary history as a result of different evolutionary pressures.

Doolittle (1993) also proposed a tentative scheme for the evolution of the vitamin K-dependent coagulation factors. First, an ancestral prothrombin emerged by serine protease gene duplication and the acquisition of Gla and EGF domains. Prothrombin's EGF domain(s) served as a site for the binding of tissue factor which at this time possessed the ability to activate it to yield thrombin. After the emergence of fibrinogen, factor X appeared as a result of a prothrombin gene duplication. The ability of factor X to activate prothrombin released the latter from its dependence on tissue factor. Factor VII, duplicated from factor X, was able to bind tissue factor an to activate factor X. Factor IX emerged last, again duplicated from factor X. Prothrombin then acquired kringle domains via exon shuffling allowing it to bind fibrin. The concomitant loss of prothrombin's EGF domains abolished its now redundant interaction with tissue factor. The plausibility of this scheme was tested by Krawczak et al. (1996) whose approach is described in subsequent sections.

Reconstruction of Mammalian Ancestral cDNAs

Krawczak et al. (1996) reconstructed the catalytic domains of the early mammalians ancestors of the vitamin K-dependent factors and the common ancestor of all five proteins. The study of Krawczak et al. (1996) relied upon mammalian phylogenies from different sources and included cDNA sequences from human, macaque, sheep, pig, rabbit, rat, mouse, dog and cow. Conserved cDNA blocks were used as anchors to align intervening non-conserved cDNA blocks for a particular protein, and since protein function must have been retained during evolution, frameshift mutations were precluded in all alignments. For each protein, the authors then deduced form the alignments the most likely cDNA sequence at each node of the mammalian phylogenies employed, including the roots representing the respective mammalian ancestors. A small number of ambiguities remained as to the ancestral mammalian cDNA sequences. These were resolved by reference to relative nearest neighbour-dependent mutation rates in humans. Use of human-derived parameters in this context was justified on the basis of the long-term evolutionary stability of relative single base-pair substitution rates.

The evolution of a gene family whose members acquire different functions is invariably accompanied by rapid amino acid sequence divergence in functionally important regions. Indeed, specific examples of this phenomenon include the 'accelerated evolution' (hypervariability) of the reactive center regions of serine protease inhibitors and the active site region of some serine proteases. For humans, the mutation rates derived during the reconstruction process were found by Krawczak et al. (1996) to be significantly higher in the factor VII, factor X and protein C lineage than in the factor IX and prothrombin lineages. Similarly in the dog, the factor VII lineage exhibited a higher mutation rate than that of factor IX. That factor IX exhibited a lower mutation rate than the other proteins was indicative of its early emergence. Once adapted to its functions, factor IX would have bad to change rather less over evolutionary time than the other proteins of more recent origin which still had to adapt to their new-found roles. For factor IX, protein C and prothrombin, mutation rate differences were also apparent between species and exhibited in inverse correlation with generation time. Consistent with previous results, the mutation rate in humans was much less than that in rodents.

Evolutionary Divergence of the Vitamin K-dependent Coagulation Factors

Krawczak et al. (1996) identified a number of highly conserved regions in their reconstructed ancestral mammalian cDNA sequences. Mismatches in these regions were classified on the basis of whether they corresponded to either a silent or a missense mutation during the process of evolutionary divergence. Interestingly, the numbers of silent and missense mutations did not correlate with each other. This finding was interpreted in terms of the existence of two distinct molecular clocks: one would be based upon silent mutations and would run constantly after the divergence of any two sequences. The other would be based upon missense mutations which would continue to run immediately after gene duplication but would stop, or at least be dramatically slowed, once the new protein product had acquired, and then become adapted to, its new biological function.

Since by far the smallest number of silent mutations in conserved codons had occurred since the divergence of factors VII and X, Krawczak et al. (1996) concluded that these two proteins must have their most recent root in common. This assertion is not inconsistent with the fact that the human factor VII (F7) and factor X (F10) genes are not only syntenic but also very closely linked on chromosome 13q34, as a result of their recent emergence through a process of duplication and divergence. Since the conserved cDNA blocks of factors IX exhibited the largest number of neutral differences with respect to other proteins, this was held to imply that factor IX was the first to diverge from the other proteins. However, a potential pitfall with this conclusion could have been a faster rate of substitutions at the X-linked factor IX (F9) locus as compared to the other, autosomal genes. However, at least for human, substitution rates in the F9 gene were significantly lower than in the factor F7, F10 and protein C (PROC) genes which also appeared to be the case in dog. Thus, divergence time rather than a higher propensity to mutate was held to be responsible for the large number of silent substitutions which separate the F9 cDNA sequence form the other cDNAs. The precise order of divergence events for protein C, prothrombin and the common ancestor of factors VII and X could not be clarified unequivocally by Krawczak et al. (1996) on the basis of their data.

The cDNA based phylogeny of the five vitamin K-dependent factors emphasizes the previously recognized relatedness of (i) protein C and prothrombin and (ii) factors VII and X. However, it is markedly different from other proposed phylogenies. This may have been because earlier attempts employed alignments of amino acid rather than cDNA sequences. The phylogeny of Krawczak et al. (1996) was closest to that presented by Doolittle and Feng (1987): both phylogenies claimed a comparatively recent common root for the catalytic domains of factors VII and X. However, the major difference lies in the much earlier branching out of the catalytic domain of factor IX, postulated by Krawczak et al. (1996).

Gene duplication has played a very important role in the evolution of the genomes of higher organisms. Indeed, there is now considerable evidence for saltatory increases in the number of genes around the time of the emergence of the vertebrates 500 Myrs ago. These increases appear to have been caused by the duplication and subsequent divergence of many different gene sequences. It is still impossible to say for certain, however, when the duplication and divergence of the vitamin K-dependent serine proteases of coagulation occurred during vertebrate evolution. Prothrombin is present tin bony fish (trout), cartilaginous fish (dogfish) and in the hagfish, one of the modern representatives of the jawless *Agnatha*. Thus, although thrombin is found in the most primitive of modern vertebrates, there is no evidence for its existence in either protochordates or echinoderms. Whether the other four vitamin K-dependent factors of coagulation are present tin these types of fish is as yet unclear. If they are, the adaptive radiation of the vitamin K-dependent factors of hemostasis must have occurred during the space of some 50 Myrs between the divergence of the protochordates and the appearance of the *Agnatha*, some 450 Myrs ago. The processes of gene duplication and divergence that have led to the emergence of the five present day vitamin K-dependent factors of coagulation provide further evidence of what must have been a very active phase in the evolution of the modern vertebrate genome.

Reconstruction of a Common Ancestor of the Vitamin K-dependent Coagulation Factors

The evolutionary scenario was based solely upon the highly conserved region of the reconstructed ancestral mammalian cDNA sequences. However, this phylogeny was used by Krawczak et al (1996) in an attempt to reconstruct the common ancestor of all five vitamin K-dependent factors by means of molecular modelling. To this end, the less well conserved regions of the mammalian ancestral sequences were first aligned in the order stipulated by the phylogeny. The alignment was then used to determine the nucleotides present at each node of the phylogeny moving upwards through the tree to its root. The resulting cDNA sequence, representing the putative common ancestor of all five vitamin K-dependent factors, contained several gaps which yielded ambiguities in the sequence of the ancestral protein. However, since the protein core had to contain a full complement of residues for meaningful molecular modelling to be possible, all such amino acids were replaced with the analogous residue found in extant human thrombin as a 'best guess'.

One way to examine the plausibility of the deduced amino acid sequence of the vitamin K-dependent factor ancestral protein would have been to express it *in vitro* and to characterize it biochemically. An alternative strategy was to construct a model of the tertiary structure of the protein by comparative methods and examine its topology and biophysical properties. Such a model was created by Krawczak et al. (1996) using the X-ray crystallographic coordinates of the heavy chain of α-thrombin as a template. This template was aligned against the high resolution structure of seven other serine proteases and the amino acid sequence of the reconstructed ancestral protein. Sequence modifications became necessary in this process at two variable regions, which were longer than the analogous loops present tin extant serine proteases, and a single unpaired Cys residue (Cys22), which was replaced by Ala as in extant human thrombin. Extant human thrombin was chosen for comparison in this and other instances since, of all of the modern hemostatic serine proteases, it bore the strongest sequence homology to the ancestral protein. Moreover, it was assumed that the vitamin-K dependent factor ancestral protein would have been capable of performing the end effector function of thrombin in the hemostatic cascades, that is the cleavage of soluble fibrinogen to generate insoluble fibrin clot.

The putative ancestral protein was found to contain 86 charged groups at pH 7.0, very similar to the number (89) found in extant human thrombin. In both cases, the great majority of these charges were noted to be accessible to water. The global electrostatic distribution across the ancestral protein's surface was relatively uniform by comparison with the dipolar distribution evident in extant human thrombin. Both structures still possessed an unbroken equatorial belt of negative charge but the extent of the electrostatic field strength in the ancient protein was not as strong as that of extant thrombin. It was speculated that the increase in charge intensity and dispersal over evolutionary time reflected a trend toward increasing protein binding specificity.

The fibrinogen-binding exosite (P1) was found to have greatly increased in size during the evolution of thrombin. From the electostatic contour map, only five Arg (126, 165, 233) and three Lys residues (230, 243, 245) of the putative ancestral protein appeared to contribute to a small patch of positive charge. Moreover, several thrombin residues known to be important in the binding of fibrinogen were not present in the ancient protein suggesting that the ancestral protease bound fibrinogen only weakly. A number of residues in thrombin have been shown to be involved in the binding of protein C and these

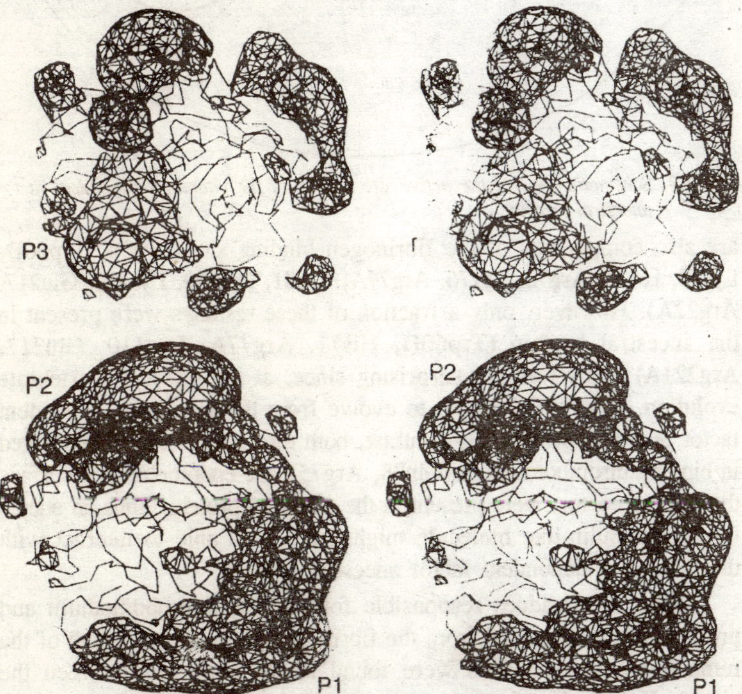

Fig. 3.2. Stereo views of the electrostatic profiles of the putative vitamin K-dependent factor ancestral protein (above) and extant human thrombin (below).

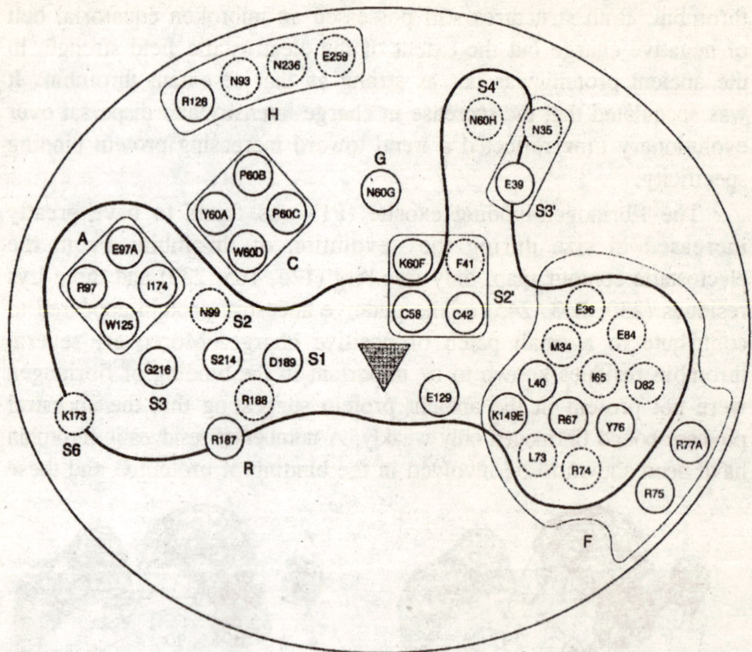

Fig. 3.3. Schematic view of the active site canyon of the vitamin-K-dependent factor ancestral protein.

are also components of the fibrinogen binding site (Lys36, Trp60D, Lys70, His71, Arg73, Tyr76, Arg77A, Lys81, Lys109, Lys110, Glu217, Arg22A). However, only a fraction of these residues were present in the ancestral protein (Trp60D, His71, Arg77A, Lys110, Glu217, Arg221A). This was not surprising since, at the dawn of vertebrate evolution, protein C had yet to evolve from the vitamin K-dependent factor ancestral protein. By contrast, both thrombin residues implicated in binding thrombomodulin (Gln38, Arg75), the endothelial cell surface thrombin receptor, were present in the ancestral protein. Thus an ancien thrombomodulin-like molecule might have been able to interact with the vitamin K-dependent factor ancestral protein.

When the residues responsible for binding thrombomodulin and protein C were excluded from the fibrinogen binding patch, five of the remaining seven residues were found to be conserved between the vitamin K-dependent factor ancestral protein and extant human thrombin (Lys60F, Asn60G, Asp186A, Lys186D, Glu192). Both of the altered residues (Thr60I, Asp222) exhibit conservative changes (to Ser and Glu, respectively). These seven residues may have constituted the original fibrinogen binding patch that was to increase in size and binding affinity as well as diversifying functionally over evolutionary time.

The extensive positive electrostatic charge associated with the heparin binding site in extant thrombin was much smaller in size and field strength in the model of the putative ancient protein. In prothrombin, the second kringle domain interacts with the heparin-binding site to slow down antithrombin III/heparin-mediated inhibition prior to proteolytic prothrombin activation. The primitive heparin binding site evident in the ancien protease would have been unable to bind the kringle 2 domain as strongly as its extant descendant. This is consistent with the view that the ancient protein contained a light chain of Gla and EGF domains, the latter only being replaced by kringles at a later stage in the evolution of the protease. The heparin-binding site may then have evolved in such a way as to balance the dual requirements of antithrombin III/heparin-mediated inhibition of thrombin and the kringle 2-mediated protection of prothrombin from premature inhibition.

Evolution of Human Protein C

One of the best characterized vitamin K-dependent serine protease of coagulation is protein C. Activated protein C (APC) exerts a negative feedback regulatory effect on both the intrinsic and extrinsic pathways of coagulation through the proteolytic inactivation of factors Va and VIIIa in the presence of protein S and a negatively charged phospholipid surface. Protein C is activated by thrombin through cleavage of the Arg169-Leu170 bond with release of a dodecapeptide from the heavy chain. This reaction is enhanced some 20,000 fold by thrombomodulin, an endothelial cell surface glycoprotein which binds thrombin with high affinity.

The substrate specificity of serine proteases is largely dependent upon the structure and properties of the substrate-binding pocket adjacent to the active site. By analogy with other trypsin-like serine proteases, Asp189 (chymotrypsin numbering) of APC forms the bottom of this binding pocket. In principle, the presence of Ser197 at the S2 substrate-binding site of APC would allow the binding pocked to accommodate larger amino acid residues such as Phe or Leu. However, this theoretical diversity appears not to be capitalised upon by the native structure since the substrate specificity of APC is in reality confined to factors Va and VIIIa.

A molecular model of the catalytic domain of early mammalian protein C was also derived and was used to examine the functional architecture of that ancient domain and to explore its evolutionary progression from the putative common ancestor of the vitamin K-dependent serine proteases toward extant human protein C. This application of homology modeling to a reconstructed amino acid sequence made it possible to trace the evolution of structural features in a human protein and, in so doing, to make certain inferences regarding the development of its functional specificity. Since its first appearance in evolution as a result of a gene duplication, the catalytic domain of human protein C has undergone 41 amino acid changes, at least some of which have presumably served to 'fine-tune' interactions with its substrates. Thus, those sites of protein-protein interaction already present in early mammalian protein C should contain the amino acid residues most essential for interaction and hence for the biological functions of the protein.

Comparison of the Functional Architecture of early Mammalian Protein C with the Putative Ancestor of all Vitamin K-dependent Factors

Six of the 13 variable regions (VR's) in the model of early mammalian protein C were derived by reference to the crystal structure of factor Xa. The similarity in VR loop length between early mammalian protein C and extant factor X is consistent with both proteins sharing a considerable period of evolutionary history. It would seem reasonable to suppose that factor X (an activator of prothrombin and therefore a procoagulant) and protein C (an anticoagulant by virtue of its inhibitory action in inactivating factors Va and VIIIa) have coevolved in such a way as to ensure that thrombin generation is appropriately promoted yet adequately limited in response to hemostatic challenge.

Doolittle (1993), Patthy (1990) and Krawczak et al. (1996) all concluded that protein C and prothrombin probably emerged at a similar stage of vertebrate evolution. In the primordial vertebrate hemostatic system, one role for protein C could have been to act as part of a negative feedback control mechanism. Since both the catalytic efficiency of thrombin and the efficiency of activation of prothrombin were presumably undergoing a continual process of optimization at that time, there would have been an increasing requirement for an efficient negative regulator of thrombin.

Any change of function from the common ancestor of the vitamin K-dependent factors to early mammalian protein C would have necessarily required changes in the active site of the protease, thereby altering its substrate specificity from fibrinogen to the emerging co-

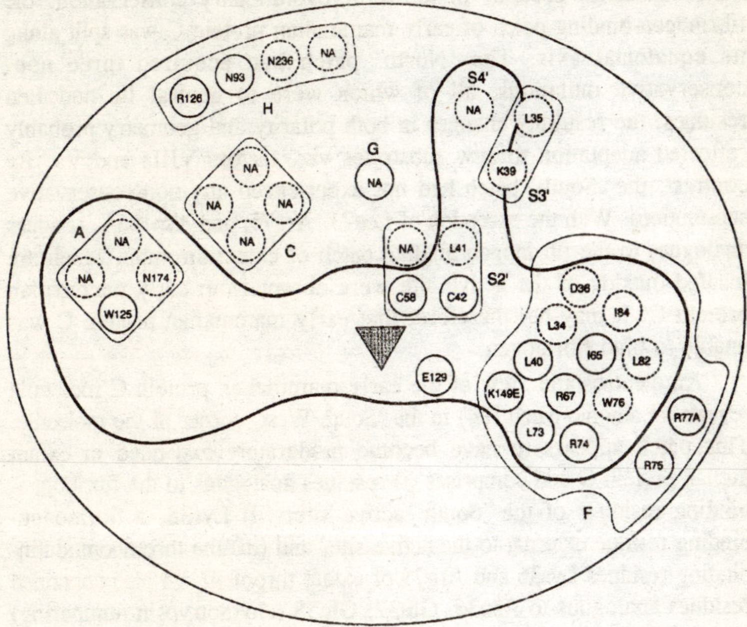

Fig. 3.4. Schematic view of the active site canyon of early mammalian protein C.

factors VIIIa and Va. The 'accelerated evolution' of active sites is thought to have been important in the diversification of the substrate specificity of serine proteases, subsequent to gene duplication. From inspection of our models, changes in the active site region of protein c appear to have involved the elimination of the aryl binding site (with the exception of residue W215 which is highly conserved between all serine proteases) and the chemotactic binding site. Both of these modifications resulted from the loss of two pincer-like highly mobile insertion loops, Leu59-Asn62 and (to a lesser extent) Leu144-Gly150, originally present in the vitamin K-dependent factor ancestral protein. Since both loops are required for the catalytic activity of extant thrombin, it may be inferred that the early mammalian ancestor of protein C had already lost its ability to cleave fibrinogen. Moreover, the absence of those residues which in thrombin bind protein C (with the exception of Leu73, Arg75, and Asp186A) implies that early mammalian protein C was probably not auto-catalytic.

In the early mammalian protein C molecule, the distribution of residues analogous to the fibrinogen-binding site of extant thrombin was very different from that predicted in the putative K-dependent factor ancestral protein. In terms of evolutionary conservation, the fibrinogen-binding patch of early mammalian protein C was split along its equatorial axis. The 'North' patch had acquired three non-conservative mutations, all of which were to neutral Ile and Leu residues; the resulting changes in both polarity and geometry probably reflected adaptation to new substrates viz. factors VIIIa and Va. By contrast, the 'South' patch had not experienced any non-conservative substitutions. With the exception of Leu73, Arg75, and Asp186A, residues analogues to the fibrinogen-binding patch of extant thrombin which are located outside of the active site were absent from early mammalian protein C. It may be, therefore, that early mammalian protein C was unable to bind fibrinogen.

An electrostatic view of the early mammalian protein C molecule reveals an anionic patch (P4) in the 'South-West' corner of the molecule. This patch appears to have become moderately expanded in extant human protein C and comprises (i) residues analogous to the fibrinogen-binding residues of the 'South' active site, (ii) Lys38, a fibrinogen-binding residue external to the active site, and (iii) the thrombomodulin-binding residues Lys36 and Arg75 of extant thrombin. P4 also contained residues analogous to Asn36, Gln37, Glu38 (chymotrypsin numbering) of vitamin K-dependent factor ancestral protein (Glu38 belongs to the S3' site of extant thrombin). In early mammalian protein C, these

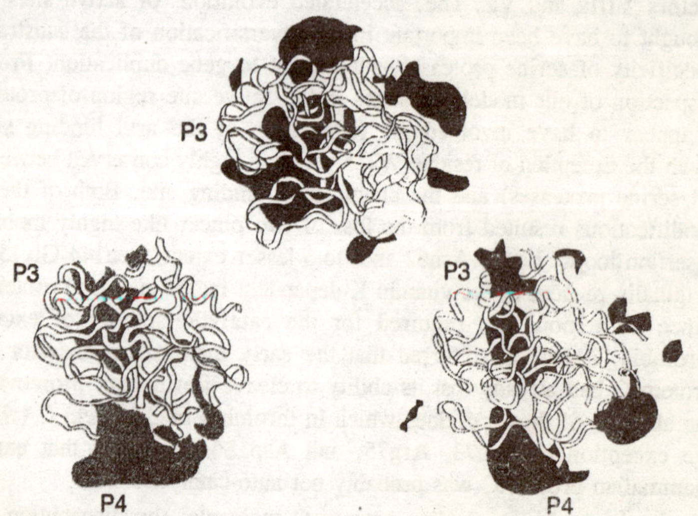

Fig. 3.5. The electrostatic profiles of the vitamin K-dependent factor ancestral protein, early mammalian protein C and extant human protein C are shown.

three residues had become lysines which would have served to increase the net positive charge in this area. Since this region binds thrombomodulin in extant protein C, the P4 site may represent the nascent thrombomodulin-binding patch in early mammalian protein C. The envisaged evolutionary transformation of the primitive original fibrinogen/thrombomodulin-binding patch into a region specialised for thrombomodulin binding. In extant protein C, substitution of Trp76 by Arg has served to increase further the anionic potential, and thus specificity of this patch. Finally, it should be noted that the anionic residues of P4 (K36, K37, K38, R74, K148, R149, R151) in early mammalian protein C appear to have undergone relatively few subsequent substitutions in order to 'fine-tune' binding to thrombomodulin. Thus the P4 patch in early mammalian protein C was probably of the minimum size required to bind thrombomodulin, and was subsequently extended and refined by evolution.

The binding patch P3 (comprising anionic residues Lys20, Arg23, Lys159, and Arg188) of the vitamin K-dependent factor ancestral protein is also apparent in early mammalian protein C. However, in the latter molecule, this patch was less expansive and confined to a higher latitude in the molecule. The apparent migration of the patch away from the equatorial belt appears to have continued during evolution toward extant human protein C. The persistence of the anionic potential of this area through evolutionary time, together with the report of a type II (dysfunctional) substitution in this region, implies a functional role in both ancient and extant proteins.

Comparative Geometry of the Active Sites of Early Mammalian and Extant Protein C

Extant human protein C and early mammalian protein C exhibit 83.6% sequence identity at the amino acid sequence level, and exhibit a backbone RMS divergence of 2.6 Å (computed against a model of the serine protease domain of extant human protein C). Recently, a relatively low resolution crystal structure of human APC (des Gla domain) has been solved. Comparison of our model with this structure served to demonstrate that the majority of functionally important residues identified in extant protein C were already present in its early mammalian predecessor. The Ca^{2+} binding loop (residues 70-80 with the exception of N78), the active site loop (residues 146-152) and the insertion helix (residue 129) are all to be found intact in early mammalian protein C. The asymmetric surface charge distribution of extant protein C is also apparent in early mammalian protein C.

Finally, residues E192 and Y143 (the major determinants of substrate specificity) which span the active site, are present in both proteins.

Since the regions close to the catalytic triad of extant human and early mammalian protein C are likely to possess a similar functional architecture (owing to the minimal divergence in the template structures underlying the two models), the active site geometries of the two proteins may be reliably contrasted at relatively high resolution. Within the catalytic site pocket of extant human protein C, the S2 residue (Ser 198) is predicted to have arisen by substitution of a Phe residue present in early mammalian protein C. This would imply that an initially tightly fitting catalytic pocket became more capacious during evolution. Why, however, the substrate specificity of extant human protein C is restricted to factors Va and VIIIa, although its active site pocket can theoretically accommodate larger substrates, remains unclear. With this exception, the amino acid sequence of the active site pocket of early mammalian protein C is identical to that of extant human protein C. Thus the structural features noted in the crystal structure of human APC would probably also have been found in the early mammalian protein.

A prominent hydrophobic and solvent-accessible region is found on the surface of the serine protease domain of early mammalian protein C. This region has been shown to bind the second (C-terminal) EGF-like domain of the light chain in factors Xa, IXa, and protein C (residues 100-116 and 45-50 respectively). Early mammalian protein C may therefore have possessed a light chain, an hypothesis consistent with Patthy's (1985) view of serine protease evolution.

In summary, the application of homology modeling to a reconstructed amino acid sequence allowed Wacey et al. (1997) to trace the evolution of specific structural features in protein C. This approach provided new insights into the evolution of protein C, allowed an assessment of the nature of the minimal thrombomodulin binding site and permitted inferences to be made as to the possible catalytic mechanism in early mammalian protein C. Such an approach lends itself readily to the study of the common ancestors of other multigene families and may therefore allow further 'structure-function studies' of phylogenetically reconstructed proteins to be performed.

4

Reconstruction of Gross Genes

Gross gene rearrangements involving the inversion, translocation or fusion of DNA sequences are relatively rare in inherited disease although much more common in cancer. Clearly such mutations can lead to dramatic changes in gene structure, function and/or expression. At first glance, it may therefore seem somewhat surprising that there are numerous examples of such mutations that have occurred during evolution. However, genes do not necessarily always change by a slow incremental process of single base-pair substitution and sudden gross changes have also played an important role in fashioning our genes.

Inversions

Most examples of inversions that have occurred during mammalian evolution have involved large portions of chromosomes and are usually *pericentric* in that the breakpoints are located on opposite chromosomal arms with the inversion spanning the centromere. This contrasts with *paracentric* inversions in which the inversion involves breaks on the same chromosomal arm, with a segment from only one chromosome are being inverted.

Pericentric Inversions

A good example of a pericentric inversion that has occurred during evolution is that of two clusters of zinc finger protein genes on human chromosome 10. Cluster A, comprising ZNF11A, ZNF25, ZNF33A, and ZNF37A, is located on chromosome 10p11.2 whilst cluster B, which is a partial duplicate of cluster A, comprises ZNF11B, ZNF33B, and ZNF37B and is located on chromosome 10q11.2. Duplicated gene clusters are usually contiguous but these ZNF clusters are located on opposite sides of the centromere consistent with their involvement in a pericentric inversion subsequent to the duplication. Tunnacliffe et al. (1993) proposed that this event took place during primate evolution.

The human genes encoding NADP-dependent malate dehydrogenase (ME1; 6q12), glutathione S-transferase 2 (GSTA2; 6p12) and phosphoglucomutase 3 (PGM3; 6p12) form part of a chromosome 9 syntenic group in mouse. The extension of this syntenic group across the centromere of chromosome 6 in human is indicative of a pericentric inversion which must have occurred since the divergence of rodents and humans. Other examples of inferred pericentric inversions that have occurred during primate evolution involve the β-casein (CSN2; 4q13-q21) gene, the GL1 oncogene on chromosome 12q13, the fibroblast growth factor genes FGF3 and FGF4 on chromosome 11q13 and the steroid sulfatase pseudogene (STSP; Yq11). Finally, a Y-specific pericentric inversion occurs as a polymorphism among Gujarati Indians.

Pericentromeric regions appear to be particularly prone to arrangement and this genetic instability could be related tot the presence of specific repetitive sequences.

Paracentric Inversions

An example of a human-specific paracentric inversion is that involving the short arm of the Y chromosome. A single contiguous segment of Xq21 is homologous to two noncontiguous segments of Yp and it is likely that the transposition of a ~4 Mb segment from the X to the Y chromosome ~3–4 Myrs ago after the divergence of human from chimpanzee was followed by an inversion of the sequence. Schwartz et al. (1998) suggested that this inversion could have been mediated by recombination between two LINE elements misaligned at homologous CATTATTCT motifs. Since humans from different racial groups have all been found to possess this Yp inversion, the rearrangement must have occurred prior to the radiation of the human racial groups.

Another rather different type of paracentric inversion is exemplified by the Xq28 inversion polymorphism involving the Emery-Dreyfuss muscular dystrophy (EMD) and filamin (FLN1) genes. Recombination between two large 11.3 kb inverted repeats (with >99% sequence homology) flanking these genes has been responsible for a frequent and clinically asymptomatic inversion of the 48 kb FLN1/EMD region. The inversion polymorphism may have resulted either from inter- or

Fig. 4.1. Structure of the Xq28 region containing the human emerin (EMD) and filamin (FLN1) gene loci.

intra-chromatid exchange between misaligned repeats. Whichever, it is very common in the general population occurring in the heterozygous state in 33% of females and 19% of males (18% of human X chromosomes surveyed).

Other regionally localized paracentric inversions may be inferred from the presence of long inverted repeat regions containing homologous but divergently transcribed genes. Examples of this include the α-amylase (AMY1A and AMY1B) and the α- and β-fibrinogen (FGB and FGG; 4q31) genes. Genes which are divergently transcribed, evolutionarily related and which share bidirectional promoters are likely to have arisen by a process of duplication and inversion. Other genes have clearly evolved by this same process but are almost certainly too far apart to share promoter elements, for example the serum amyloid A1 and A2 genes (SAA1, SAA2; 15-20 kb apart on chromosome 11p15) and the GABA receptor β3 and α5 genes (GABRB3, GABRA5; 100 kb apart on chromosome 15q11-q12).

Physiological and Pathological Inversions

Somatic inversions, both relatively small and rather larger involving megabase-sized DNA fragments, occur physiologically during rearrangement of DNA at the immunoglobulin k (IGKV; 2p12) and T cell receptor β (TCRB; 7q35) loci in human. In human pathology, sporadic chromosomal inversions are not uncommon although each type of inversion is likely to be individually rare. By contrast, intragenic DNA sequence inversions, occurring as a result of recombination between inverted repeats in the germline are highly unusual. The best known example is that found in the factor VIII (F8C) gene causing hemophilia A: this rearrangement occurs in about 40% of severely affected patients and recurs at high frequency. The mechanism responsible is thought to be homologous intrachromosomal recombination between a gene (F8A) located in intron 22 of the F8C gene and one of two additional homologues of the F8A gene situated 500 kb upstream of the F8C gene.

Intragenic Inversions

Not surprisingly, intragenic inversions occurring during gene evolution are extremely uncommon. One example is however provided by the family of inter-α-trypsin inhibitors encoded by four genes in the human genome (ITIH1, ITIH3, ITIH4, 3p21; ITIH2, 10p14-p15). The ancestral ITIH gene was first duplicated with one copy being translocated to chromosome 10 about 300 Myrs ago. Prior to further gene duplication and divergence, the primordial ITIH1/ITIH3/ITIH4 gene experienced an inversion of exons 3-13.

Common Sites for Inversions in Pathology and Evolution?

It has been known for some time that the chromosomal inversions in human pathology are nonrandomly distributed although it remains unclear whether this is due to interchromosomal differences or to bias to ascertainment. Some of these inversions have nevertheless been characterized with sufficient precision to allow comparison with inversions that have occurred in the karyotypic evolution of the hominids. Miro et al. (1992) performed such a comparison and demonstrated that 10/20 pericentric inversions and 1/4 paracentric inversions which had occurred during human chromosome evolution coincided (albeit at low level resolution) with known sites of pathological inversion. These (evolutionary) inversions were inv(1) (p12q21.22), inv(2) (p11.2q13), inv(4) (p14q21.1), inv(5) (p13.3q13.3), inv(7) (q11.23q22.2), inv(8) (p21.1q22.1), inv(9) (p24.2q12), inv(11) (p15.5q13), inv(16) (p11.2q12.1), inv(18) (p11.32q11.2), and inv(Y) (p11.2q11.23). If the sites that have been involved in chromosome inversions during primate evolution were also to be involved in cases of human chromosome pathology, this would argue for certain chromosomal regions possessing sequence characteristics (possibly long highly conserved inverted repeats) that could predispose to this type of lesion. Thus higher resolution studies such as that on human chromosome 12q15 (which contains breakpoints associated with both benign solid tumors and a pericentic inversion that occurred during hominoid evolution) could yield valuable insights into the mechanisms underlying chromosome rearrangement in both pathology and evolution.

TRANSLOCATIONS AND TRANSPOSITIONS

The human genome is replete with examples of the transposition and translocation of gene sequences during evolution. A selection of some of the most notable examples are given here while others have been discussed in the context of gene duplication. The occurrence of most translocations has been inferred from the localization of evolutionarily related genes on different chromosomes or different chromosomal arms. Another is the human γ-aminobutyric acid receptor

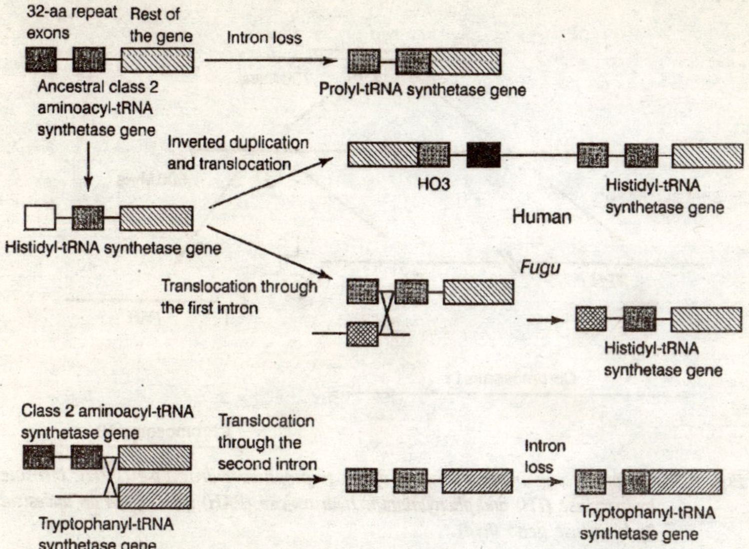

Fig. 4.2. Hypothetical translocation events in the evolution of the aminoacyl-tRNA synthetase genes.

($GABA_AR$) which comprises several different types of subunit that combine to form a pentameric channel complex and which are encoded by a small but dispersed gene family. The GABR1, GABRB2, and GABRG2, genes have been localized to chromosome 5q34-q35 whilst similar clusters comprising GABRA2, GABRB1, and GABRG1 (4p13-p12) and GABRA5, GABRB3 and GABRG3 (15p11.2-q12) are present on two other chromosomes. This organization is compatible with the duplication of an ancestral gene cluster and its translocation to other chromosomes.

Tryptophan hydroxylase, tyrosine hydroxylase and phenylalanine hydroxylase are members of the family of pterin-dependent aromatic amino acid hydroxylases and are encoded by the TPH, (11p15.3-p14), TH (11p15.5) and PAH (12q22-q24.1) genes, respectively. PAH and TPH are estimated to have arisen by a process of duplication and divergence some 750 Myrs ago whilst PAH and TH diverged subsequently about 600 Myrs ago. The synteny of the extant TPH and TH genes implies that the PAH gene must have been translocated to chromosome 12 only after the second duplicational event 600 Myrs ago. A similar explanation may pertain for the insulin (INS; 11p15.5), insulin-like growth factor 1 (IGF1; 12q22-q24.1) and insulin-like growth factor 2 (IGF2; 11p15.5) genes.

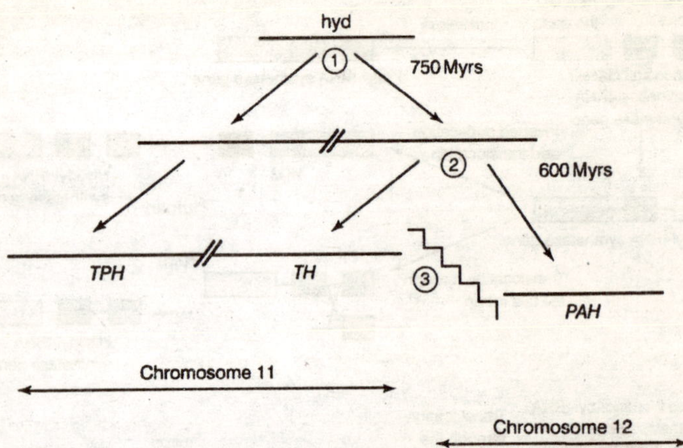

Fig. 4.3. Genetic events in the evolution of the tryptophan hydroxylase (TPH), tyrosine hydroxylase (TH) and phenylalanine hydroxylase (PAH) genes from an ancestral hydroxylase gene (hyd).

In humans and orangutans, two $tRNA^{Asn}$ gene clusters are located on the short and long arms of chromosome 1, respectively (TRN, 1p36.1; TRNL, 1q21). By contrast, Old World monkeys possess only one $tRNA^{Asn}$ gene cluster on chromosome 1p whilst the capuchin (a New World monkey) possesses a single cluster on 1q. These data are consistent with the interpretation that the $tRNA^{Asn}$ gene cluster split before the divergence of Old World monkeys and hominoids about 30 Myrs ago. No simple inversion can account for this situation and, since this chromosome is well conserved in primate evolution, Buckland et al. (1992) speculated that one $tRNA^{Asn}$ gene cluster could have been relocated from one chromosome arm to the other by some form of replicative transposition.

Another example of a gene translocation occurring during the evolution of the primates is provided by the adenine nucleotide translocator 3 (ANT3) and steroid sulfatase (STS) genes. The ANT3 and STS genes are pseudoautosomal (Xp22.32) in humans and other higher primates but both localize to an autosome in lemurs suggesting that there was an autosome-to-X/Y translocation after the simians diverged from the prosimians.

Pericentromeric-directed Transposition

Several example of pericentromeric-directed transposition have been described in primate genomes and since these have been well characterized, they are worthy of discussion in some detail here. Comparative FISH analysis of chromosomes from various primates has shown that a 27 kb region of Xq28 containing the creatine transporter (SLC6A8) gene and five exons of the 'CDM' gene (DXS1357E) has been duplicated in its entirety and translocated to 16p11.1, probably within the last 10 Myrs. An inverted cluster of *Alu* repeat and a number of immunoglobulin-like CAGGG repeat motifs lie in the vicinity of the Xq28 breakpoint. Either type of repeat could have mediated the transposition event but the CAGGG repeats may be the more likely candidate for involvement since they have been found to flank a number of other recently duplicated gene sequences. The chromosome 16 SLC6A8 paralogue (SLC6A10) retained both its putative promoter and transcriptional activity but the chromosome 16 CDM paralogue clearly represents a truncated pseudogene. A very similar Xq-28-16p11 transposition, reported by Eichler et al. (1997), involves the independent transposition of a 9.7 kb segment encompassing exons 7-10 of the adrenoleukodystrophy (ALD; Xq28) gene to the pericentromeric regions of chromosomes 2p11, 10p11, and 16p11, and 22q11). This ALD paralogy domain lies only 27 kb telomeric to the SLC6A8/CDM paralogy domain. The autosomal sequences represent truncated non-processed pseudogenes with sequence divergence being consistent with the duplication/transposition events having occurred between 5 Myrs and 10 Myrs ago. As far as the ALD transposition events are concerned, an initial event directed to chromosome 2p11 appears to have served as a 'seed' for further pericentromeric-directed transposition events to occur. An inverted cluster of *Alu* repeats lies in the vicinity of the Xq28 breakpoint. In addition, Eichler et al. (1997) noted the presence of a GCTTTTTGC repeat flanking the duplicated region which they speculated might serve as a sequence-specific integration site for transposition. Such a sequence may provide a hyper-recombinogenic signal which could account for the propensity of this locus to be involved in transpositional events.

Other pericentromeric-directed duplications/transpositions in the human genome involve the immunoglobulin Vk light chain locus, the immunoglobulin heavy chain V_H region, the GABA A receptor α5 (GABRA5) gene and the neurofibromatosis type 1 (NF1) gene. The inter-chromosomal duplication and transposition of NF1-related sequences may be related to the presence in the vicinity of α-satellite sequenced that could have served to promoter their dispersal. The transposition associated with the immunoglobulin Vk light chain locus (IGKV; chromosome 2p11.1→chromosome 1) was directed to a site containing an ALD like GCTTTTTGC repeat suggesting the a similar mechanism may have bene responsible for these transpositional events.

The pericentromeric zinc finger gene cluster on chromosome 19p12 which harbors the ZNF208 gene is flanked by large blocks of β-satellite repeat sequences. This gene cluster is thought to have arisen early in primate evolution (~50 Myrs ago) by a process of pericentromeric-directed transposition. Eichler et al. (1998b) proposed a model in which an ancestral ZNF gene became associated with β-satellite repeat sequences at 19p12. Such repeats are capable of rapid expansion, possibly by unequal crossing over, and may have served to promote the rapid amplification of the associated ZNF gene.

The above example serve to indicate that the pericentromeric regions of human chromosomes have frequently acquired sequences from remote genome locations. Indeed, these regions appear to be very dynamic, being subject to amplification, duplication, deletion and inversion event as well as translocations. Many of these pericentromeric rearrangements have occurred relatively recently during primate evolution leading to marked inter-specific differences even among the great apes. The potential evolutionary importance of pericentromeric regions can perhaps be gauged from Eichler's (1999) description of them as 'recruitment stations for repeats' and 'reservoirs for the accumulation of transposed genic segments'.

Sub-telomeric Transposition

Sub-telomeric regions may be similarly dynamic. Thus the rapid proliferation of multigene families clustered near telomeres may have occurred by repeat-mediated bursts of duplication/transposition of short stretches of genomic DNA (e.g. the olfactory receptor genes). Duplicational transposition also resulted in the telomeric localization of a number of pseudogenes derived from the *Chl1*-related helicase (DDX11, 12p11; DDX12, 12p13) and interleukin 9 receptor (IL9R; Xq28/Yq12) genes. The 'spreading' of a sub-telomeric region has also been described my Monfouilloux et al., (1998) originally localized to 17 qter in chimpanzee and orangutan, a specific sub-telomeric domain has been translocated in humans and has colonized several other chromosome ends. Finally, there is emerging evidence for sequence exchange between human telomeric and centromeric regions. This may have had consequences for the growth and distribution of multigene families as in the case of the human olfactory receptor genes which, although clustered predominantly in subtelomeric regions, are also located in pericentromeric regions.

Some sequence exchanges that have been reported as translocations may, however, upon closer scrutiny be explicable by other mechanisms.

The 'inter-chromosomal polymorphism' involving a sub-telomeric exchange between region on chromosomes 4q35 and 10q26, which has been shown to be present in at least 20% of the human population, is a case in point. These regions are thought to be 100-400 kb in length and contain 3.3 kb tandem repeat units (each containing two homeodomain sequences and two different classes of GC-rich repetitive DNA) that are 98% homologous to each other. An explanation invoking frequent-inter chromosomal translocation would be virtually unprecedented. The most parsimonious explanation would appear to be inter-chromosomal gene conversion followed by nonhomologous recombination leading to repeat homogenization.

Translocations and Chromosome Associations

Do some chromosomal rearrangements occur as a consequence of the spatial association of particular chromosomes on the metaphase plate? Nagele et al. (1995) have suggested that human chromosomes are arranged in antiparallel fashion as a rosette and that this chromosome arrangement is both consistent and cell-type independent inhuman cells. The high frequency of translocations involving certain chromosomes, whether in pathology (somatic/germline) or evolution, may therefore be due at least in part to certain chromosomes being positioned in close proximity to each other during the cell cycle.

Translocations in Pathology and Evolution

As with inversions, the chromosomal sites that have been involved in translocations during mammalian evolution sometimes crop up in cases of human chromosome pathology. Thus the breakpoint of X; 1 papillary renal cell carcinoma-associated translocations has been mapped to a <200 kb region between the SPTA1 and CD1C genes at the same 1q21 location that contains the boundary between human and mouse syntenic regions. Another example of such a correspondence is provided by the holoprosencephaly critical region at 2p21, a site which is associated with a translocation breakpoint in a gibbon karyotype. If such correspondence are not simply coincidental, these chromosomal regions could possess characteristics that would predispose to translocations. Their study could therefore yield valuable insights into the mechanisms underlying chromosome rearrangement in both pathology and evolution.

Gene Fusion

Gene fusions resulting from somatic chromosomal translocations are a common cause of tumorigenesis. By contrast, gene fusions occurring in the germline and causing genetic disease (e.g. hemoglobin

Lepore) are a highly unusual cause of human genome pathology. Such fusions are however thought to have had an important role in evolution through the creation of novel genes encoding novel combinations of functional protein domains; some examples of this phenomenon are described below. Exon shuffling has been an important process in evolution, serving to bring together coding sequence blocks form different sources to generate new proteins capable of novel interactions and therefore, potentially, with novel functions. In a sense, therefore, all proteins have evolved by a series of gene fusion event. In this section, however, we concentrate solely on the generation of novel genes through the fusion of once distinct and independently functional gene sequences by recombination.

Gene Fusion during Evolution

Carboxyesterase E1, an enzyme responsible for the detoxification of ingested xenobiotics, exhibits sequence homology both with acetylcholinesterase and the carboxy terminal end of thyroglobulin, a precursor of thyroid hormone. Phylogenetic analysis has suggested that both acetylcholinesterase (human ACHE gene on 7q22) and thyroglobulin (human TG gene on 8q24.2-q24.3) evolved from a common ancestral gene that encoded a carboxyesterase and that the emergence of thyroglobulin must have preceded the divergence of the vertebrates and invertebrates. The evolutionary origin of the amino terminal portion of extant thyroglobulin is however unknown.

The human multidrug resistance (PGY1; 7q21.1) gene encodes a membrane-associated pump protein called P-glycoprotein. Sequence homology between the amino and carboxyl terminal halves of the protein was initially held to be consistent with the view that the protein evolved by duplication of a primordial gene. However, once the structure of the 29 exon PGY1 gene had been determined, it became clear that only two intron pairs (both within nucleotide binding domains) were located in conserved positions in the two halves of the protein. Thus, rather than a primordial duplication, Chen et al. (1990) proposed that primordial proteins corresponding to the left and right halves of P-glycoprotien were formed independently by the fusion of closely related genes encoding the nucleotide binding domain with genes for different transmembrane domains. Subsequent fusion of these two independently derived genes then resulted in the formation of the PGY1 gene. Pauly et al. (1995) have speculated that a highly conserved poly(CA).poly(TG) sequence in intron 15 of the PGY1 gene could have, with *Alu* sequences in introns 14 and 17, mediated such a fusion event by recombination.

The human glutaminyl-tRNA synthetase (EPRS) gene serves to encode two distinct aminoacyl-tRNA synthetase activities joined as part of a multi-enzyme synthetase complex. The EPRS gene, located at chromosome 1q41-q42, comprise 29 exons spanning some 90 kb. Exons 4 to 10 encode a glutaminyl-tRNA synthetase whilst exon 19 to 29 encode a prolyl-tRNA synthetase. The function of the intervening region (exons 11 to 18) is unclear. Since the glutaminyl- and prolyl-tRNA synthetases belong to different enzyme classes which are believed to have evolved by separate pathways, the composite EPRS gene would appear to represent the consequence of an ancient gene fusion.

The human quiescin Q6 (QSCN6; 1q24))gene also appears to represent the product of an ancient gene fusion that occurred during metazoan evolution. The N-terminal portion of the quiescin Q6 protein is related to the thioredoxin family (human members include thioredoxin (TXN; 9q31) and prolyl-4- hydroxylase β-polypeptide (P4HB; 17q25) whilst the C-terminal is related to the yeast *Erv1* growth regulatory gene (human homologue, GFER, 16p13).

The human lamin B receptor is an integral protein of the inner nuclear membrane encoded by a gene (LBR) on chromosome 1q42.1. The LBR gene comprises 13 protein coding exons, the first four of which encode the N-terminal domain and the remainder the C-terminal sterol reductase-like domain. A relatively large intron separating exons 4 and 5 is suggestive of an ancient recombination between two genes, one encoding a basic nuclear protein, the other a sterol reductase. Consistent with this interpretation, the 3' end of the LBR gene is structurally homologous with two other human genes viz. the transmembrane protein TM7SF2 (TM7SF2; 11q13) and 7-dehydrocholesterol reductase (DHCR7; 11q13).

A highly unusual example of a gene fusion is provided by the human HLA-DRB6 (6p21.3) gene. This gene has represented something of a paradox in that although both exon 1 and the promoter region are absent in orthologous human, chimpanzee and rhesus macaque genes, these genes are still capable of being transcribed. This apparent paradox has now been resolved by the elucidation of the molecular basis of the change: the insertion of a retroviral (mouse mammary tumor virus) LTR into intron 1 of the primate HLA-DRB6 gene >23 Myrs ago, prior to the divergence of the Old World monkeys form the human-ape lineage. Whether the exon/promoter deletion accompanied or instead followed the LTR insertion is unclear. What is clear is that an open reading frame for a new exon was created by the insertion which serendipitously encoded a hydrophobic sequence that was able to function as a leader for the truncated HLA-DRB6 protein. The new exon provided a functional donor splice site at its 3' end which potentiated in-register splicing with exon 2 of the HLA-DRB6 gene. The LTR also provided a substitute promoter region essential for the transcription of the downstream gene. Thus, this case not only provides an example of the *de novo* creation of an exon but also illustrates a potential evolutionary mechanism to retarget proteins within the cell. The incorporation of a novel leader peptide into, for example, a cytoplasmic protein that previously lacked one could result either in an alteration in the cellular localization of the protein or its conversion into a secreted protein.

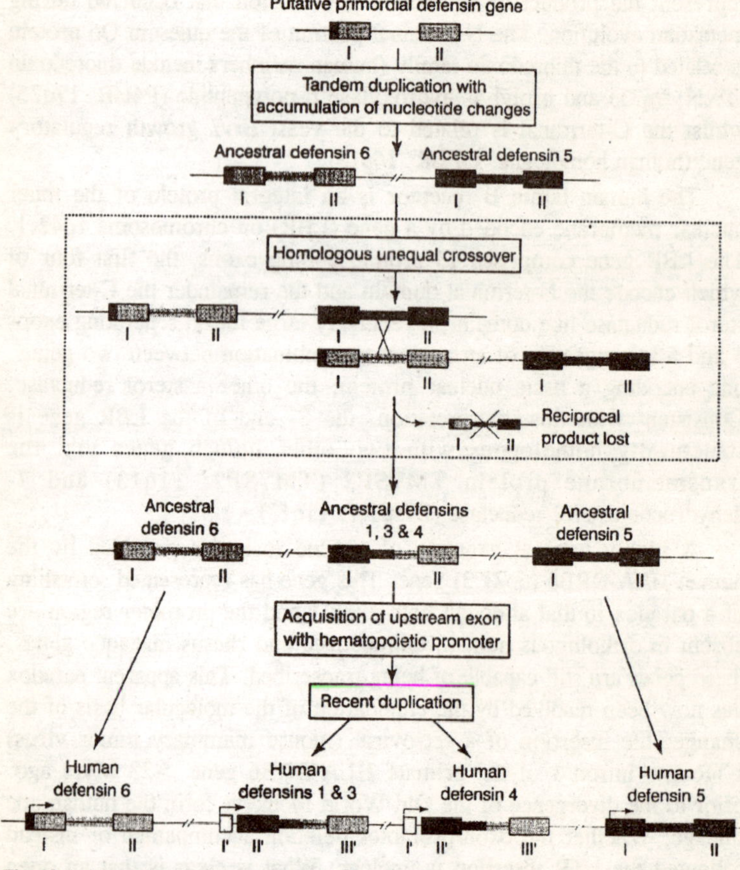

Fig. 4.4. Model for the involvement of an homologous unequal crossover in the evolution of the human defensin gene family.

The mammalian definitions constitute a family of microbicidal and cytotoxic peptides made by neutrophils. In humans, they are encoded by a gene cluster (DEFA1, DEFA3, DEFA4, DEFA5, DEFA6, DEFB1) at chromosome 8p23. After a primordial defensin gene had duplicated to yield two genes ancestral to extant DEFA5 and DEFA6, an unequal crossing over event generated a novel hybrid defensin gene which was the ancestor of the present day hematopoietic defensin genes DEF1, DEFA3, DEFA4.

The SCYA18 (17q11.2) gene, which encodes a number of the small inducible cytokine family, appears to have been generated by the fusion of two macrophage inflammatory protein-1α-like (SCYA3) genes with subsequent deletion and selective use of exon. Since there are several related genes (SCYA3, SCYA3L1, and SCYA3L2) in the vicinity of SCYA18, the authors suggested that the SCYA3 gene might represent a 'hot spring' that continually generates new genes by duplication and fusion. The human immunoglobulin γFC receptor IIC (FCGR2C; 1q23) gene is also thought to have resulted from an unequal crossover, this time between the FCGR2A and FCGR2B genes (1q23); the 5' end of the FCGR2C gene exhibits significant homology to the FCGR2B gene whilst its 3' end is homologous to the FCGR2A gene. Finally, two human ubiquitin genes represent the products of an in-frame fusion event between a ubiquitin gene and a ribosomal protein gene encoding a protein of either 52 (UBA52; 19p13) or 76-80 (UBA80) amino acids in length.

Internal Methionines as Evidence for Ancient Gene Fusion Events

Berman et al. (1994) demonstrated that some 20% of eukaryotic protein are multiples of 123 ± 3 amino acids. This modular structure has been held to reflect the combinatorial fusion of genes, initially of the same elementary size, during evolution. One consequence of this fusion process would be the frequent occurrence of methionine-encoding ATG triplets at the borders between unit length sequence segments, a prediction which has been born out by statistical analysis. This positional preference of methionines testifies to the excision-reinsertion mechanism of protein construction and means that these internal methionines can justifiably be termed the 'fossils of gene fusion.' Further, it illustrates the probable importance of gene fusion events in the evolutionary construction of extant gene sequences.

Fusion Gene Polymorphisms

Some fusion genes occur in the human genome as polymorphic variants. One example is the red and green visual pigment genes (RCP,

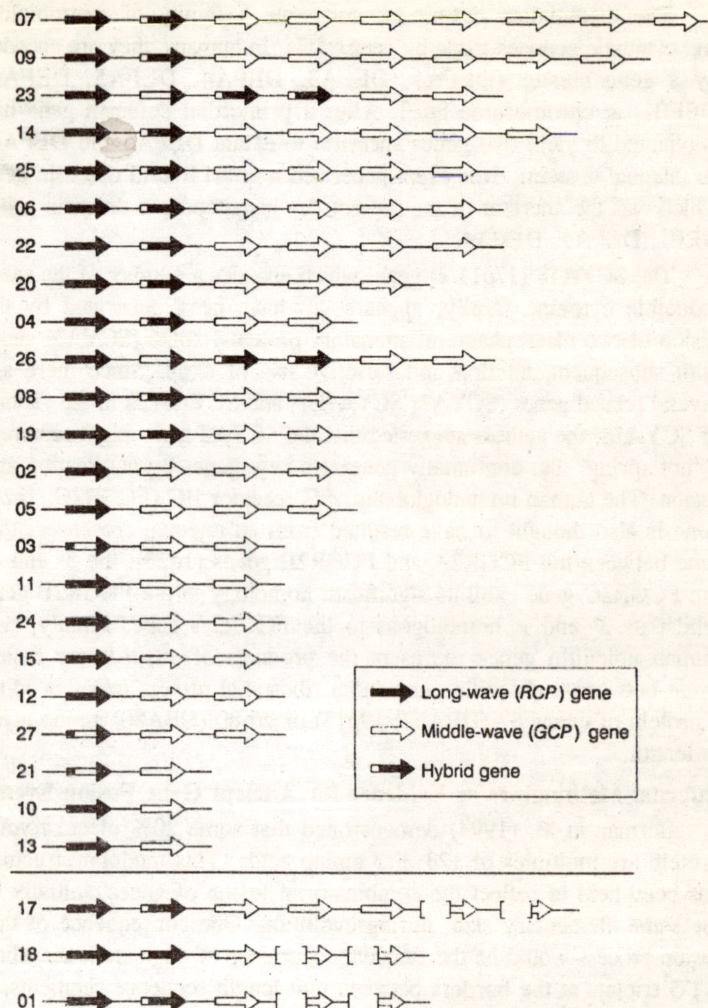

Fig. 4.5. Inferred visual pigment gene (RCP, GCP) arrangements in 26 individuals with normal colour vision.

GCP; Xq28). Normal human males with trichromatic vision typically posses one RCP gene, one, two or more GCP genes plus and RCP/ GCP hybrid gene. In some individuals, the RCP genes contain a substantial amount of GCP sequence e.g. exon 4 of the GCP gene. Other individuals possess RCP genes that contain 5' sequences derived from GCP genes. The degree of polymorphism is dramatic in that ~70% of individuals with normal trichromatic colour vision possess one or other type of fusion gene. Other gene fusions have occurred as a result of unequal homologous crossing over to generate polymorphic variants in the MNS (GYPA; 4q31) and ABO blood group systems (ABO; 9q).

Fusion Splicing

Although gene fusion can arise through deletion or translocation, it need not invariably occur as a result of DNA rearrangement. Gene fusion can, in functional terms, also be brought about by the *fusion splicing* of mRNA transcripts derived from two closely linked but unrelated genes. Evidence for such a mechanism has come from the cotranscription of two human genes encoding galactose-1-phosphate-uridylyl-transferase (GALT) and interleukin-11 receptor α-chain (IL11RA) which are closely linked (separated by only 4 kb) on chromosome 9p13. GALT is a 43 kDa enzyme required for the conversion of galactose to glucose and is in coded by a gene which comprises 11 exon spacing 4 kb and which specifies a 1.4 kb mRNA. The IL11RA gene, a member of the hematopoietin receptor superfamily, comprises 13 exons spanning 8 kb and encodes a 48 kDa protein. Magrangeas et al. (1998) demonstrated that the two genes are sometimes cotranscribed in normal human cells and that the 3 kb fusion mRNA encodes an 85 kDa protein of unknown function and biological significance. The in-frame fusion transcript probably results from a combination of inefficient RNA polymerase II termination of GALT gene transcription and an alternative splicing event between exon 10 of the GALT gene and exon 2 of the IL11RA gene.

Magrangeas et al. (1998) speculated that fusion splicing could represent an 'exploratory event for evolution'; fusion proteins thus formed might initially be produced at low cellular levels but would represent a reservoir of novel functions which could be called upon to confer a selective advantage under particular conditions. In such a situation, it may be envisaged that a point mutation at the acceptor splice site or polyadenylation site could then lead to the generation of much larger amounts of the fusion protein. Such a mutation is evident in the guinea pig gene that encodes seminal vesicle secretory proteins 1, 3, and 4. The 5' half of this guinea pig gene is homologous to the human semenogelin II (SEMG2; 20q12-q13) gene. Indeed, sequences related to the human SEMG2 gene are also found in the first intron of the guinea pig gene. However, the 3' half of the guinea pig gene shares homology with the closely linked skin-derived antileukoproteinase/elafin (PI3; 20q12-q13) gene. It would appear that, as a result of an

A→G transition, the guinea pig gene contains a novel AG dinucleotide 7 bp upstream of the 3' splice site used in the SEMG2 gene. If used as a splice acceptor site, this AG dinucleotide would lead to a spliced product that is out of frame with the product of the SEMG2 gene, and the use of a splice acceptor site in the downstream PI3 gene would have been favoured. Thus an alternative splicing pathway has led to the creation of a novel guinea pig gene that presumably encodes a protein with a new function.

In principle, fusion splicing could also lead to the creation of novel genes through the germline retrotransposition of the fusion mRNA. Such novel fusion genes would however be expected to be characterized by the absence of introns (unlike the base of the guinea pig seminal vesicle secretory protein gene cited above) and to be chromosomally distant from the parental genes. There are at least two other possible ways of generating novel genes by fusion splicing. The first would be a via *trans*-splicing (involving and joining of independently transcribed coding sequences from genetically unlinked loci) followed by retrotransposition back into the genome. As yet, however, the available evidence for *trans*-splicing in mammalian cells must be regarded as somewhat tentative. The second mechanism would involve the retrotransposition of abnormally spliced gene transcripts comprising exons joined in an order different from that in which they are normally found in the genome (*scrambled exons*). Exon scrambling has been described for at least two human genes: DCC (18q21) and MLL (11q23).

Recombination

Numerous recombination events have how been documented as having occurred during the evolution of mammalian genes. Indeed, most gene duplication and amplification events have been mediated by recombination. Examples of the emergence of novel human gene sequences by recombination-mediated duplication include those encoding saposin C (PSAP; 10q), pepsinogen A3 (PGA; 11q13), mannose-binding protein (MBP; 10q11-q21) and the α-amylase genes (AMY1A, AMY1B, AMY1C, AMY2A, AMY2B, 1p21). Recombination events have also been invoked to explain cases of gene fusion.

Multiple independent recombination events are thought to have occurred during the evolution of the two human haptoglobin genes (HP and HPR; 16q22.1) and their counterparts in the other primates. Whilst the great apes and Old World monkeys possess three haptoglobin genes (Hp, Hpr, and Hpp), New World monkeys only have one (Hp). This is consistent with a gene triplication after the divergence of Old World from New World monkeys followed by a Hpp gene deletion in the human lineage. Breakpoint analysis suggests that both duplications and deletions of gene copies may have been mediated by homologous unequal recombination between *Alu* repeats. That this region is prone to recombination is also evidenced by the haptoglobin gene copy number polymorphism found in the black population. Other examples of homologous recombination events thought to be responsible for human genes copy number polymorphisms include the α-globin (HBA1) genes, the ζ-globin (HBZ) gene, the α-amylase genes, the proline-rich protein genes (PRB1, PRB2, PRB3, PRB4; 12p13.2), and the pepsinogen A genes (PGA3, PGA4, PGA5; 11q13); none of these copy number polymorphisms are known to have any clinical significance

Recombination has also resulted in the alteration of the expression level of a gene. Thus, the galago δ-globin gene is expressed at an unusually high level in the adult (18% of β-like globin chains, cf. 0-6% in other primates) as a result of a recombination event which replaced 2.4 kb of δ-globin gene sequence by β-globin gene sequence containing 800 bp of the promoter region.

Considerable effort has been made to localize and characterize the DNA sequences responsible for mediating the recombinational events known to have occurred during gene evolution. Thus, the duplication events involved in the evolution of the human glycophorin (GYPA, GYPB, and GYPE; 4q28.2-q31.1) genes, the human growth hormone gene (GH1, GH2, CSH1, CSH2; 17q22-q24) cluster and the mouse lysozyme genes are considered to have been mediated by recombination between *Alu* repeats. A correspondence between illegitimate recombination junctions and the sites of *Alu* sequence insertion has also been proposed for the primate α-globin (HBA1, HBA 2; 16p13.3) genes. The duplication of the primate γ-globin (HBG1, 11p15.5) gene may have been mediated by homologous recombination between LINE elements flanking a fetal globin progenitor gene. High G+C content may promote recombination whilst Ashley et al. (1993) have suggested that telomeric repeats could promoter meiotic recombination.

High resolution genetic mapping studies have provided evidence for both 'hotspots' and 'coldspots' or recombination. A frequent disease-associated recombinational hotspot in the human genome occurs at 17p11.2-p12 where a CMT1A-REP tandem repeat mediates meiotic crossing over events through chromosome misalignment. Humans and chimpanzees have two copies of this repeat whilst gorilla, orangutan and gibbon have only a single copy. The CMT1A-REP repeat must

1. Duplication

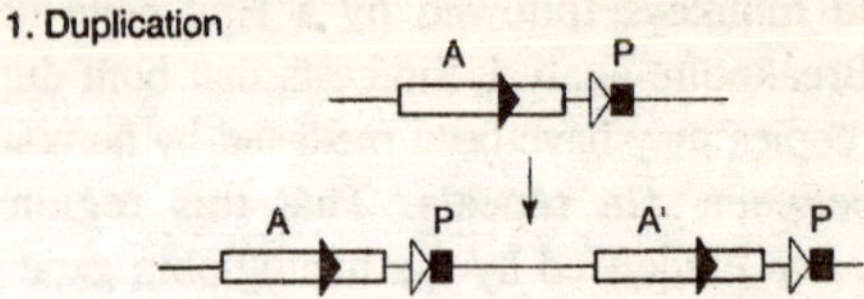

2. Unequal homologous *Alu-Alu* recombination

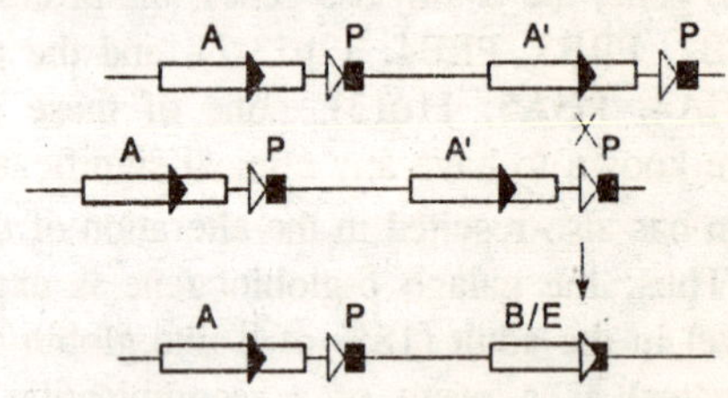

3. Duplication

A P B/E

A P B E

Fig. 4.6. Three-step scenario in the evolution of the glycophorin gene family.

therefor have appeared before the divergence of chimpanzee and human. The repeat contains a *mariner*-like element in the vicinity of the recombination hotspot and, since this element occurs in association with the CMT1A-REP repeat in all primates, it must have predated the emergence of the proximal and distal copies of the repeat in the human-chimpanzee common ancestor. Interestingly, the *mariner* element of *Drosophila* exhibits sequence homologies to transposons from *Caenorhabditis elegans* as well as heptamer and nonamer signal sequences of the vertebrate immunoglobulin somatic recombination pathway. Whether these homologies that transcend the invertebrate-vertebrate divide are merely coincidental, or whether they reflect either a common evolutionary origin or alternatively convergent evolution, is as yet unclear.

Recombination may have been involved in at least some instance of 'exon shuffling' the process of intron-mediated recombination by which functional domains encoded by one or more exons are dispersed to variety of different proteins. For example, a combination of X-ray crystallographic data and studies of exon organization have led to the conclusion that the two eons encoding the nucleotide-binding domain have been independently transferred to the human phosphoglycerate kinase (PGK1; Xq13.3) gene as well as to the maize alcohol dehydrogenase and chicken glyceraldehyde-3-phosphate dehydrogenase genes. Whilst the almost ubiquitous presence of introns may permit exon shuffling by intron-mediated recombination, it has been suggested that the presence of short unconserved introns with different types of splice junction may serve to prevent recombination between exon clusters encoding highly conserved functional domains without inhibiting the potential for domain (multiexon) shuffling.

V(D)J Recombination

In the lymphocyte, functional immunoglobulin and T-cell receptor genes are assembled from their constituent gene coding segments by V(D)J recombination. Recombination is directed by recombination signal sequences (RSS) adjacent to each coding sequence. The RSS comprises conserved heptamer and non-amer motifs separated by a non-conserved spacer region of either 12 bp or 23 bp (termed 12 signals and 23 signals); recombination only occurs between a 12 signal and a 23 signal. V(D)J recombination is initiated by the products of two lymphoid-specific intronless recombination-inactivating genes, RAG1 and RAG2, closely linked to each other on human chromosome 11p13. The RAG1 and RAG2 proteins bind the two recombination signals, bringing them into close juxtaposition and cleaving the DNA molecule thereby separating the recombination signals from the flanking coding segments. The RAG1 and RAG2 genes are present in all vertebrates but have not been reported either in jawless vertebrates or invertebrates.

Together RAG1 and RAG2 constitute a transposase that is capable of excising DNA segment containing recombination signals from a donor site and inserting it into a target DNA molecule. The product of transposition contains a short target site duplication immediately flanking the transposed fragment that is reminiscent both of retroviral integration and the insertional mechanisms employed by other transposases. It is therefore reasonable to speculate that the RAG1 and RAG2 genes may once have been contained within a transposable element that also possessed flanking RSS sequences. Since in extant genomes, the RAG genes and the RSS elements are unlinked, transposition now serves to relocate pairs of RSS elements plus the intervening sequence without the necessity of the RAG genes themselves being displaced.

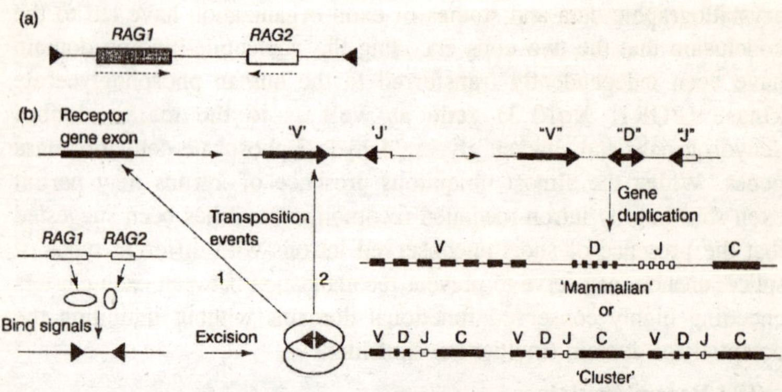

Fig. 4.7. Model for the evolution of the immunoglobulin and T-cell receptor (TCR) genes and the role of RAG-mediated transposition.

The split nature of the vertebrate immunoglobulin light chain and T-cell receptor α and γ genes could have originated from the germline insertion of the RAG transposon into an ancestral receptor gene soon after the divergence of jawed and jawless fishes. This gene could then only have been expressed if the inserted transposon were excised by the RAG proteins and the two end of the exon rejoined and repaired. The tripartite structure characteristic of the immunoglobulin heavy chain and T-cell receptor β and γ genes could have arisen as a result of the further insertion of a second RAG transposon into the same exon. Subsequent duplication of the individual gene segments or of the entire gene would then have served to generate the 'mammalian-type' or 'cluster-type' organization of mammals and cartilaginous fishes, respectively. It therefore appears likely that our early vertebrate ancestors were successful in taming a transposon by transforming it into a site-specific recombinase that was then harnessed in the cause of generating the genetic diversity so vital for the flexible adaptive response of our immune system.

Gene Conversion

Gene conversion is the 'modification of one of two alleles by the other'. The end result is very similar to that resulting from a double unequal crossing-over event and in humans, it is difficult to distinguish the two processes. However, in practical terms , the difference is that the correction of an 'acceptor' gene or DNA sequence by gene conversion is nonreciprocal leaving the 'donor' sequence physically unchanged.

The process of gene conversion may involve the whole or only a part of a gene and usually occurs between highly homologous, but nonallelic genes. Gene conversion may be suspected in cases where the degree of sequence homogeneity exhibited by related genes is much greater than expected from what is known of their evolutionary history. Thus, two or more gene sequences may exhibit strong homology over all or parts of their coding or promoter sequences yet they are known from phylogenetic studies to have diverged a considerable time ago and might reasonably have been expected to have accumulated significant numbers of both synonymous and non-synonymous substitutions.

There are now numerous examples of gene conversion causing human gene pathology but gene conversion has also had an important influence in evolution. Examples of gene conversion in human gene evolution often involve duplicated gene sequences lying in close physical proximity: the G-γ (HBG2; 11p15.5) and A-γ (HBG1; 11p15.5) globin genes, the β- and δ-globin genes (HBB, HBD; 11p15.5), the α1-(HBA1) and α2-(HBA2) globin genes, the HLA-DQB1 loci (6p2.3), the visual pigment (GCP and RCP; Xq28) genes, the rhesus blood group (RHD, RHCE; 1p34-p36.2) genes, the salivary (AMY1A, AMY1B, AMY1C; 1p21) and pancreatic (AMY2A and AMY2B; 1p21) amylase genes, the

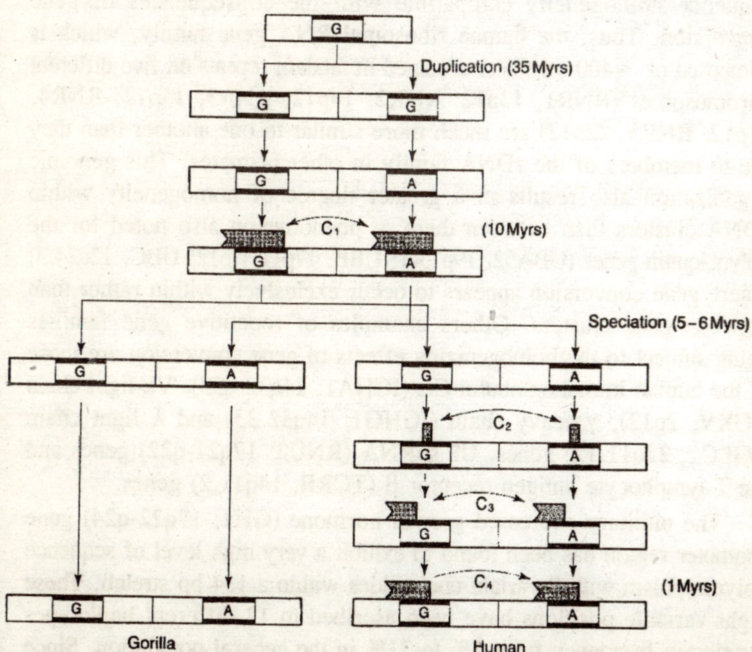

Fig. 4.8. History of γ-globin gene evolution and conversion events in the great apes.

α-interferon genes IFNA1 and IFNA13 (9p22), the haptoglobin genes (HP, HPR; 16q22.1), the p15 and p16 cyclin-dependent kinase inhibitor (CDKN2B and CDKN2A; 9p21) genes, the glycophorin genes GYPA and GYPE (4q28.2-q31.1), the P glycoprotien 1 and 3 genes (PGY1, PGY3; 7q21.1), the salivary proline-rich protein genes (PRB1, PRB2, PRB3, PRB4; 12p13.2), the cardiac α- and β-myosin heavy chain (MYH6, MYH7; 14q12) genes, the chorionic somatomammotropin (CSH1, CSH2; 17q22-q24) genes, the immunoglobulin $C_H\alpha$ (IGHA1, IGHA2, 14q32) and V_H genes (IGHV, 14q32), and the α1-acid glycoprotein (ORM1, ORM2; 9p34.1-34.3) genes.

In the context of gene conversion, if there is any middle ground between pathology and evolution, it is probably occupied by the cytochrome P450 CYP2A6 gene (19q13.2). A null CYP2A6 variant, formed by gene conversion between the CYP2A6 gene and the closely linked but inactive CYP2A7 gene occurs at polymorphic frequencies in the Japanese (28%) and African-American (2.5%) populations.

Presumably, the greater the number of homologous repetitive elements, the greater the opportunity for gene conversion. Consistent with this assertion, multigene families frequently exhibit extensive sequence homogeneity compatible with the consequences of gene conversion. Thus, the human ribosomal RNA gene family, which is composed of ~400 members arranged in tandem repeats on five different chromosomes (RNR1, 13p12; RNR2, 14p12; RNR3, 15p12; RNR4, 21p12; RNR5, 22p12) are much more similar to one another than they are to members of the rDNA family in other primates. This genomic organization also results in a greater degree of homogeneity within rDNA clusters than between them, a phenomenon also noted for the polyubiquitin genes (UBA52, 19p13.1; UBB, 17p11.1-p12; UBC, 12q24.3) where gene conversion appears to occur exclusively within rather than between gene clusters. Others examples of repetitive gene families being subject to the homogenizing effects of gene conversion are those of the human immunoglobulin Cα (IGHA1; 14q32-q33), Vk light chain (IGKV; 2p12), γ heavy chain (IGHG1; 14q32.33) and λ light chain (IGLC1; 22q11.12) genes, U2 snRNA (RNU2; 17q21-q22) genes and the T-lymphocyte antigen receptor β (TCRB; 14q11.2) genes.

The pituitary-expressed growth hormone (GH1; 17q22-q24) gene promoter region has been found to exhibit a very high level of sequence polymorphism with 8 variant nucleotides within a 134 bp stretch. These eight variable positions have been ascribed to 12 different haplotypes ranging in frequency from 2% to 31% in the general population. Since they occur in the same positions in which the GH1 gene differs from the other placentally expressed genes of the growth hormone cluster [two chorionic somatomammotropin (CSH1 and CSH2) genes, a chorionic somatomammotropin 'pseudogene' (CSHL1) and a second growth hormone gene (GH2)], the mechanism responsible is likely to be gene conversion with the placentally expressed genes serving as donors of the converted sequences. Various examples of gene conversion of functional genes templated by pseudogenes have been documented as a cause of human pathology.

Studies of gene conversion in fungi have shown that gene conversion occurs not only between alleles but also intrachromosomally between duplicated sequences on either the same chromatid or the sister chromatid and even between sequences on nonhomologous chromosomes. Does gene conversion in the human genome occur predominantly within a single chromosome (i.e. through sister chromatid exchanges) or instead between either homologous or nonhemolytic chromosomes? Data from both the human ribosomal RNA and U2 snRNA gene families have suggested that intra-chromosomal events are more frequent than inter-chromosomal events. Inter-chromosomal gene conversion events have been reported in human gene pathology but these are comparatively rare. In an evolutionary context, possible examples of gene conversion operating between homologous loci on nonhomologous chromosomes are provided by non-reciprocal exchanges between the X-linked and autosomal phosphoglycerate kinase (PGK1, Xq13; PGK2; chromosome 19) genes and the X-linked and autosomal pyruvate dehydrogenase (PDHA1, Xp22.1-p22.2; PDHA2; 4q22-q23) genes.

Although gene conversion is generally considered to involve only short stretches of DNA sequence, it has often not been possible to determine the precise length of the converted sequence owing to the high degree of homology manifested by the genes involved. Papadakis and Patrinos (1999) reviewed data from the human Gγ- (HBG2; 11p15.5) and Aγ-(HBG1; 11p15.5) globin genes and concluded that the length of the converted fragments is usually less than 400 bp but can vary from as little as 113 bp to as much as 2266 bp.

The mechanism underlying gene conversion remains elusive but must presumably entail the close physical interaction between homologous DNA sequences. It may involve heteroduplex formation followed by mismatch repair. Both Amor et al. (1988) and Matsuno et al. (1992) have noted the presence of Chi-like sequences (GCTGGGG; known to promote recombination both in *E. coli* and in mouse immunoglobulin genes) in the vicinity of regions of the CYP21/CYP21P and HBB genes. These authors speculated that the Chi-like sequences might play a role in gene conversion events. Various other sequences e.g. *Alu* repeat sequences, a retroviral Long Terminal Repeat, alternating purine.pyrimidine tracts and $(CT)_n.(GA)_n$ microsatellite have also been postulated to be involved in promoting gene conversion. Many gene conversion event involving the HBG1 and HBG2 genes appear to terminate near a polypyrimidine stretch. Papadakis and Patrinos (1999) suggested that palindromic sequences may be associated with the termination of gene conversion, possibly through secondary structure formation blocking branch migration. The insertion of some transposable elements into members of multigene families can however reduce the rate, and limit the extent of, gene conversion by reducing the degree of homology between potential donor and acceptor sequences.

Gene conversion has thus had an important influence on the evolution of multigene families by homogenizing the sequences of duplicated or repeated genes. Sequence homogenization can of course hinder divergence and hence potential adaptation. New substitutions occurring in a sequence will tend to be lost because they are likely to be converted back to the sequence of the more common allele in the population. However, gene conversion may also promote diversity by introducing multiple sequence changes simultaneously into different members of large gene families as with the genes encoding the proteins of the major histocompatibility complex (HLA-B, HLA-DRA; 6p21.3).

Gene Creation by Retrotransposition

During the evolution of the mammalian genome, new genes have often emerged by duplication and divergence. Another, albeit less common route has been via retrotransposition. At least five functional human genes have a structure consistent with their having arisen by retrotransposition: the phosphoglycerate kinase 2 (PGK2; 19) gene, the pyruvate dehydrogenase α2 (PDHA2; 4q22-q23) gene, the cAMP-dependent protein kinase Cγ subunit (PRKACG; 9q13) gene, the SR splicing factor SRp46 (Srp46; 11q22) gene and the Y-encoded zinc finger protein (ZFY: Yp11.3) gene. All are characterized by a lack of introns and a chromosomal location different and distinct from their presumed parent genes (PGK1, Xq13; PDHA1, Xp22.1-p22.2; PRKACA, 19p13.1; ZFX, Xp21.3-p22.2 respectively). In the case of the PRKACG gene, there are also remnants of flanking direct repeats and a poly(A) tail providing further evidence of the retrotranspositional origin of the gene. This type of sequence has sometimes been referred to as a 'functional retropseudogene' but this is surely a misnomer. Functional it may be, but pseudogene it is not, and so a better term would be 'retrotransposed gene'.

The retrotransposed PGK2 gene originated prior to the divergence of eutherian and metatherian mammals some 125 Myrs ago and has been characterized in some detail. It is expressed only in spermatogenic cells whereas its putative parent (PGK1) is ubiquitously expressed. Sequence comparison studies are consistent with the newly retrotransposed PGK2 gene having possessed regulatory sequences derived form the PGK1 gene which facilitated the initial expression of the intronless gene. Inclusion of promoter sequence could have come about if the PGK1-derived PGK2 mRNA intermediate had initiated aberrantly at an upstream start site as McCarrey, (1990) suggested, or instead from an alternative upstream promoter. This promoter sequence might then be envisaged to have evolved cell type specificity, a process presumably involving loss of the CpG island still apparent in the PGK1 gene. Interestingly, the PDHA2 and PRKACG genes are also expressed exclusively in the spermatogenic cells of the testis. In the case of the X-linked PGK1 and PDHA1 genes, transposition to an autosome might have been selectively advantageous in that it would have ensured expression of these important enzyme encoding genes in Y-bearing as well as X-bearing spermatozoa.

5

GENE FLUCTUATIONS

There are now many pathological examples of the deletion, insertion, duplication and expansion of human genes causing inherited disease. Similar mutations have however also occurred over evolutionary time. Far from being invariably disadvantageous, such mutational changes have often been recruited by the opportunistic evolutionary process and now contribute to both gene and genome architecture. These types of mutation have led to significant changes in gene size and number in different lineages and their contribution to the evolution of extant human genes will now be reviewed.

GROSS GENE DELETIONS IN EVOLUTION

Gross Gene Deletions during Primate Evolution

Gross gene deletions may arise through a number of different recombinational mechanisms but probably the most common is likely to be *homologous unequal recombination* (occurring either between related gene sequences or between repetitive elements). Thus, *Alu* sequences flanking deletion breakpoints have been noted in a considerable number of human genetic conditions and may represent favoured sites for recombination and hotspots for gene deletion. Chromosomally duplicated regions (*duplicons*) are often common sites for pathological rearrangements, particularly gross deletions, since they have the potential to mediate homologous unequal recombination events (e.g. 15q11-q13). Not surprisingly, several instances of gross gene deletion have been noted during primate evolution. One such example is the loss of the γ1-globin gene in New World monkeys (with the notable exception of the capuchin monkey, *Cebus albifrons*) due to a 1.8 kb deletion which has removed most of exon 2, all of intron 2, exon 3, and much of the 3' flanking region. As a result, γ2-globin is the primary fetally expressed globin gene in New World monkeys whereas in Old World monkeys, it is γ1-globin.

Another example of gross gene deletion in primates is the loss of one of the haptoglobin (HPP; 16q22.2) genes in humans which occurred, probably via homologous unequal recombination, after the separation of the human and chimpanzee lineages. Finally, although there are two semenogelin genes (SEMG1, SEMG2; 20q12-q13.1) in humans and most old world and New World monkeys, the *Semg2* gene has been deleted from the genome of the cotton-top tamarin (*Saguinus oedipus*). There is no evidence for any selective advantage resulting from any of these gene deletions. In all cases cited, a similar paralogous gene was available to substitute for the deleted locus. This genetic redundancy probably ensured that, owing to the absence of purifying selection, such deletions came to be fixed through genetic drift.

Gross Deletional Polymorphisms

Gross gene deletions in two distinct glutathione S-transferase genes have been found to occur as polymorphic variants in various human populations. In humans, four gene families of glutathione S-transferases encode a series of enzymes responsible for the metabolism of a wide range of xenobiotics. The gene for the mu-class glutathione S-transferase (GSTM1; 1p13.3) is absent from between 10% and 64% of individuals depending upon the population under study. Loss of the GSTM1 gene is due to a 15 kb deletion probably brought about by homologous unequal recombination between two almost identical 4.2 kb repeats that flank the GSTM1 gene. These repeats are likely to have originated with the original duplication that gave rise to the mu-class glutathione S-transferase genes more than 20 Myrs ago. A similar deletional polymorphism also occurs in the theta-class glutathione S-transferase (GSTT1; 22q11.2) gene; the gene is absent in ~38% of the Caucasian population. Since the GSTM1 and GSTT1 genes are involved in

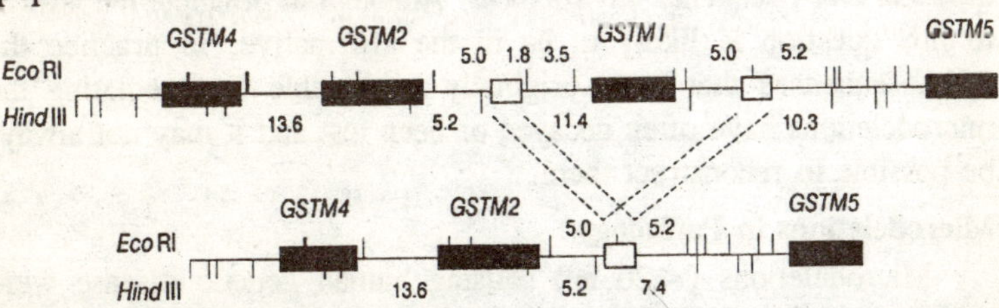

Fig. 5.1. Model for homologous recombination between 4.2 kb repeats (open boxes) flanking the human mu-class glutathione S-transferase (GSTM1) gene leading to gene deletion.

xenobiotic metabolism, it is quite possible that the presence/absence of these polymorphic alleles are not selectively neutral.

Dichromacy (colour blindness) is very common (occurring at a polymorphic frequency (5-8%) in Caucasian males) with individuals so affected having either a deletion of the green colour pigment (GCP) gene or possessing a hybrid red/green colour pigment (RCP/GCP) gene in its place. Further variable gene number polymorphisms, which probably also arose by homologous unequal recombination, are apparent in the human α-amylase gene cluster at chromosome 1p21 and the pepsinogen A gene cluster at 11q13. In the first case, a 'short' haplotype lacks AMY1A, AMY1B and the pseudogene AMYP1 whilst a 'long' haplotype contain two extra copies of duplicated fragment containing AMY1A, AMY1B, and AMYP1. In the second case, the three most common haplotypes were PGA-A (containing the PAG3, PGA4, and PGA5 genes), PGA-B (containing the PGA3 and PGA4 genes) and PGA-C (containing only the PGA4 gene). Copy number polymorphisms due to gene deletions have also been reported at the human T-cell receptor β (TCRB; 7q35) and γ (TCRG; 7p15) loci, the ζ-globin (HBZ; 16p13.3) gene, the rhesus blood group D antigen (RHD; 1p34–36.2) gene, the immunoglobulin heavy chain constant region γ4 (IGHG4; 14q32) gene and the complement C4A (C4A; 6p21.3) and C4B (C4B; 6p21.3) genes. In some gene clusters, it can be difficult to ascertain whether polymorphic alleles have arisen by gene deletion or duplication.

Microdeletions in Evolution

It is possible to extrapolate from lessons learned through the study of microdeletions in a pathological context to microdeletions that have occurred during gene evolution? In particular, can we gain insight into the nature of the generative mechanism(s) underlying evolutionarily significant microdeletions and the possible influence of the local DNA sequence environment? Although in principle the answer to this question is likely to be in the affirmative, in practice the DNA sequences that were originally responsible for mediating the microdeletions have often decayed or been lost and it may not always be possible to reconstruct them.

Microdeletions in Pathology

Microdeletions (<20 bp) causing human genetic disease were analyzed by Cooper and Krawczak (1993) in an attempt to relate the presence of specific DNA sequence motifs in the vicinity of these lesions to possible mechanisms responsible for their generation. In many cases, slipped mispairing at the replication fork between homologous sequences in close proximity to one another on complementary DNA strands appeared to be the causative mechanism. Slipped mispairing probably occurred either between *direct repeats* or through the formation of secondary structure intermediates potentiated by the presence of *inverted repeats* or *symmetric elements*. A consensus sequence, TGRRKM, common to pathological deletion hotspots has been noted in a number of different human genes. This *deletion hotspot consensus sequence* is similar to the core motifs, TGGCG and TGAGC, found in immunoglobulin switch ($S\mu$) region and to putative arrest sites for DNA polymerase α. Cooper and Krawczak (1993) also found that a second motif (polypyrimidine runs of at least 5 bp; YYYYY) was over-represented in the vicinity of short human gene deletions whilst Monnat et al. (1992) observed a significant association between HPRT1 (Xq26.1) gene deletion breakpoints and CTY vertebrate topoisomerase I cleavage sites. In principle, such sequence motifs may also have promoted the occurrence of microdeletions during evolution. Indeed, in probably the large study of its kind to date, a sequence comparison of orthologous and paralogous members of the primate T-cell receptor β (TCRB; 7q35) gene family, micro-deletion breakpoints appear to be frequently flanked by polypyrimidine runs and sequences which possess marked homology to the deletion hotspot consensus sequence.

Microdeletions Mediated by Direct Repeats

Micro-deletions occur during gene evolution at a frequency 10% that of nucleotide substitutions. Pairwise comparisons of the noncoding regions of human, rabbit and murine β-globin genes have shown that they differ from each other in terms of numerous deletions/insertions and Efstratiadis et al. (1980) proposed that the short direct repeat (2-8 bp) sequences immediately flanking these sites could have templated the generation of these lesions by slipped mispairing. Direct repeats may also have been involved in generating the two inactivating single base-pair deletions (del C822 and del G904) noted in the human α-1,3 galactosyltransferase (GGTA1; 9q34) gene. Since chimpanzees possess both these deletions, whereas orangutan and gorilla only have del904, it would appear that del904 was the original inactivating mutation. C822 is immediately flanked by imperfect direct repeats (TACAGGC<u>C</u>T and TACAAGGCAG, where <u>C</u> is nucleotide 822) that could have templated the 1 bp deletion via slipped mispairing. Slipped mispairing may also have been responsible for the G904 deletion since G904 is the 3' most base of a string of five Gs.

5' FLANKING	
Human δ	CCA-------GCATAAAA
Human β	CCAGGGCTGGGCATAAAA
LARGE INTRON	
Human A_{γ}	ATGACTTTT----ATTAGAT
Human A_{γ}	ATGACTTTTCTTTATTAGAT
Human A_{γ}	TGTGTGTGTGTGTG--------------------TGTGTGTGTGTG
Human G_{γ}	TGTGTGTGTGTGTGTGTGTGTGCGCGCGTGTGTTTGTGTGTGTGTG
Rabbit β	AAGTA------------------CTTTCTCTAATC
Mouse β	AGTCCTTCTCTCTCTCCTCTCTCTTTCTCTAATC
Rabbit β	TGGTAG-----------------------------------AAACAACT
Mouse β	TTGGCTTTTATGCCAGGGTGACAGGGGAAGAATATATTTTACATAT
Mouse β^{maj}	TGACATAGG---ATTCT
Mouse β^{min}	TGTCATAGAATAATTCTTTTTTATTTTTTATTTATTATTTTTTTCATAGAATAATTCT
Mouse β^{maj}	TGTGTGTGG------------AGTGTT
Mouse β^{min}	GGTGTGTGGATGTGAATTGTGAGTGTT
3' NONCODING	
Human ε	CAGGT------------GTTCCT
Human β	CCAATTTCTATTAAAGGTTCCT
Human A_{γ}	GCAATACAAA---TAATAAA
Human β	GTCCAACTACTAAACTGGGGGATATTATGAAGGGCCTTGAGCATCTGGATTCTGCCTAATAAA
Human A_{γ}	ATACAA---TAATAAA
Human δ	TATTTTCTGAACTTGGGAACAATGAATACTTCAAGGGTATGGCTTCTGCCTAATAAA
Rabbit β	AAAAATTAT--GGGGACA
Human β	AATTTCTATTAAAGGTTAATTTGTTCCCTAAGTCCAACTACTAAACTGGGGGATA

Fig. 5.2. Examples of deletions flanked by short direct repeats within non-coding sequences of mammalian β-globin genes.

A direct repeat may also have templated the single base deletion in the 5' flanking region of the human interferon α10 (IFNA10; 9p22) gene relative to the other α-interferon genes. Flanking direct AGGT repeats appear to have mediated an AGG deletion exhibited by both orthologous and paralogous members of the primate T-cell receptor β (TCRB; 7q35) gene family whilst in the same gene family, overlapping 7 bp direct repeats (CTTTTCTTTTCT) may have served to template a TTTCT deletion.

Microdeletions Mediated by Inverted Repeats

Inverted repeats may also have mediated the generation of microdeletions during gene evolution. One example is the inactivating 13 bp deletion in exon 2 of the gibbon urate oxidase gene; two imperfect inverted repeats (CAAGAAC and GTTCATG) span the breakpoints of this deletion. A 20 bp deletion has been reported from the 5' region of the δ-globin (Hbd) gene of the colobus monkey, *Colobus polykomos*. This deletion, which spans the transcriptional initiation site used in Old World monkeys and anthropoid apes, is responsible for a five-fold reduction in *Hbd* gene transcription as assessed by *in vitro* transcription assay. Inspection of the putative deleted bases and the flanking DNA sequence reveals the presence of a 13 bp imperfect

```
                                 +1
C aagggagggcagag--------------------CTTCTGA
R     g  a     a ccaactgttgcttATACTTG
B     g  g     a tcaactgttgcttACATTTG
S     c  g     g ctaactgttgcttTGACTTG
H     c  g     a tcgactgttgcttACACTTT
```

Fig. 5.3. Alignment of δ-globin (Hbd) gene sequences from colobus monkey (C), rhesus macaque (R), baboon (B), spider monkey (S) and human (H) showing the location of a 20 bp deletion in the colobus monkey.

inverted repeat which could have been responsible for the deletion through formation of a hairpin loop.

An inverted repeat also appears to have templated the deletion of a GAT codon in the human interferon α2 (IFNA2; 9p22) gene relative to the other α-interferon genes. Finally a contiguous inverted repeat sequence (ATTC-CCAGTTTCTGGGAAT) may well have templated an 8 bp deletion exhibited by both orthologous and paralogous members of the primate T-cell receptor β (TCRB; 7q35) gene family.

Microdeletions in Vertebrate Evolution

The comparison of gene/protein sequences between humans and the great apes also yields examples of *in-frame* microdeletions that must have occurred during primate evolution. Thus, amino acid residue Glu9 of the blue cone pigment protein (BCP; 7q31-q35) present in the talapoin monkey *Miopithecus talapoin* (an Old World primate) and in the marmoset *Callithrix jacchus* (a New World primate) is absent from the human protein and appears to have been deleted from the BCP gene within the human lineage. The functional consequences of the removal of this amino acid residue are however unclear.

Some gene region harbor a disproportionate number of deletions/insertions inferred from alignment gaps noted in sequence comparisons e.g. exons 6 of the orthologous amelogenin (AMELX, Xp22.1-q22.3; AMELY, Yp11.2) genes of various vertebrates. These lesions have dramatically reduced the similarly between vertebrate. These lesions have dramatically reduced the similarity between vertebrate amelogenins in the Pro/Gln-rich region of the protein as compared with that manifested by other region. Some microdeletions occurring during evolution may have been advantageous by virtue of their alteration of a protein product, others through a change in the reading frame bringing about gene inactivation. Of course, micro-deletions need not necessarily have conferred any selective advantage; even if merely neutral with respect to fitness, they could have become fixed by genetic drift alone.

Microinsertions in Evolution

Microinsertions that have occurred during evolution have scarcely been studied. However, the underlying generative mechanisms are likely to be broadly similar to those causing human genetic disease. In their study of microinsertions in human genes causing inherited disease, Cooper and Krawczak (1991) concluded that insertional mutation involving the introduction of <10 bp DNA sequence into a gene coding region was not a random process and appeared to be highly dependent upon the local DNA sequence context. Further, mechanistic models which have explanatory value in the context of gene deletions were found to be useful in accounting for the nature and location of gene insertions.

In noncoding DNA, insertions are about half as frequent as deletions and mostly involve single nucleotides. The rate of gap formation, regardless of whether caused by insertions or deletions, has been estimated to be ~0.15-0.17 kb^{-1} $Myrs^{-1}$. In practice, studies of the DNA sequence environment of microinsertions that have occurred during evolution are likely to be rather difficult since the original sequence context of the insertion or deletion will often have become obscured by subsequent mutation.

Gene Coding Region Microinsertions

Microinsertions occurring during human gene evolution can be found by sequence comparison of either orthologous or paralogous genes/ proteins. Thus, the 54 bp insertion in the promoter of the human liver arginase (ARG1; 6q22.3-q23.1) gene is absent in the orthologous gene of *Macaca fascicularis*. Similarly, a 37 bp insertion has been introduced into the promoter of the orthologous Duchenne muscular dystrophy (DMD; Xp21) gene of the spider monkey, *Ateles geoffroy*; the inserted sequence is flanked by two TAAA repeats. A 12 bp insertion in the T-cell receptor α-chain (TCRA; 14q11.2) gene encodes an Ile-Pro-Ala-Asp tetrapeptide (residues 88-91) that is specific to the primate lineage. Interestingly, the region of the T-cell receptor α-chain protein between positions 86 and 91 appears to be a hotspot for insertional events during evolution: the rabbit and rat genes appear to have acquired a 3 bp (single amino acid) insertion, whereas the bovine, ovine, and murine genes manifest a 6 bp (double amino acid) insertion at this position. Finally, a 24 bp sequence found in the transmembrane domain region of the human glycophorin E (GYPE; 4q28-q31) gene appears to have been derived from the paralogous glycophorin B (GYPB; 4q28-q31) gene during primate evolution prior to the divergence of the gorilla from the lineage of the other great apes. It is unclear whether this insertion event was mediated by homologous unequal recombination or gene conversion.

Microinsertion Polymorphisms

Several microinsertion polymorphisms have been reported in human genes. Thus, a single nucleotide insertion polymorphism has been noted in the promoter region of the insulin promoter factor 1 (IPF1; 13q12.1) gene. A single nucleotide insertion polymorphism is also present in the ABO blood group (ABO; 9q34) gene which serves to inactivate it. Finally, a 9 bp insertion polymorphism in exon 9 P450 CYP2D6 (22q13.1) gene occurs in the Japanese population and is associated with a poor metabolizer phenotype

Indels

Clearly, in extant proteins, selection must have ensured the retention of essential features of tertiary structure. Indeed, insertions or deletions which altered the reading frame must have been rendered harmless in order for the protein to retain its biological activity. Thus, the insertion of bases inferred in one member of a paralogous protein pair often implies a counterbalancing deletion in the immediate vicinity (or *vice versa*) to restore the reading frame. Such 'indels' tend to involve sequences of between 1 bp and 5 bp in length. They are generally found in turn and coil structures and rarely interrupt α-helices and strands. One example of a simple indel occurring during evolution is the deletion of an AA doublet and its replacement with a GT doublet in the 5' flanking region of the human interferon "α9" gene.

An example of a more complex indel that has occurred during vertebrate evolution is provided by the human chorionic gonadotropin β-subunit (CGB; 19q13.3) gene and is responsible for the introduction of a 24 amino acid C-terminal extension to the protein product. The CGB gene emerged as a result of the duplication of an ancestral β-luteinizing hormone-like (LHB; 19q13.3) gene. A single base deletion (A1540) altered the translational reading frame of the ancestral gene allowing read-through into what was originally 3' untranslated region (UTR). An insertion of a CG dinucleotide also occurred at nucleotide 1612 within the ancestral 3' UTR region and served to extend the chorionic gonadotropin β-subunit protein by a further 8 amino acids. The CGB gene therefore evolved from its LHB-like ancestor by acquiring 8 amino acids through translation in a new reading frame and incorporating 24 novel amino acids from the 3' UTR into its coding sequence.

INSERTION OF TRANSPOSABLE ELEMENTS IN EVOLUTION

Transposable elements in the human genome are essentially of two kinds, those that undergo transposition through a DNA intermediate. Transposable elements with a DNA intermediate are termed *transposons* and these are characterized by terminal inverted repeats and duplication of the target site (visible as direct repeats flanking the element). However, the great majority of transposable elements in the human genome have undergone retrotransposition through an RNA intermediate. Such *retroelements* or *retroposons* may be of either viral or nonviral origin. Whilst endogenous retroviral sequences comprise some 0.1-0.6% of the human genome, the nonviral *Alu* sequences and LINE elements may together make up as much as 10% of the genome.

Endogenous Retroviral Sequences and Transposable Elements

Retroposons

A number of retroposon families have been characterized in primate genomes. Occasionally, these are human-specific (e.g., the HERV-K10 related SINE-R.C2) but usually they are found distributed through the genomes of other primate species (e.g. RTVL-H, RTVL-I, HERV-K, and HERV-L. Type I and II HERV-H elements were amplified to ~1000 copies after the divergence of New World from Old World monkeys but before the divergence of apes from Old World monkeys. By contrast, the family of type Ia HERV-H elements expanded to ~100 copies only after the divergence of apes from Old World monkeys. Analysis of the copy number, distribution and sequence characteristics of such endogenous elements promises to provide important clues as to the evolutionary history and phylogeny of the various mammalian orders, suborders, and species, and even the population genetics of human racial groups.

Retroposons have sometimes become integrated into the vicinity of human genes. Thus, two copies of an RTV_L-I sequence are present in the human haptoglobin (HP; 16q22.1) gene cluster whilst an additional copy has been inserted in the same region in the orthologous chimpanzee gene. Rather more dramatically, the endogenous retrovirus, HRES-1 lies within the coding sequence of the human transaldolase gene (TALDO1; 11p15).

One view of endogenous retroviral elements is that they inserted themselves into the germline of our primate ancestors during the last 40 Myrs as a result of infection with exogenous retroviruses, persisting thereafter as proviruses, albeit rendered replication-defective by multiple

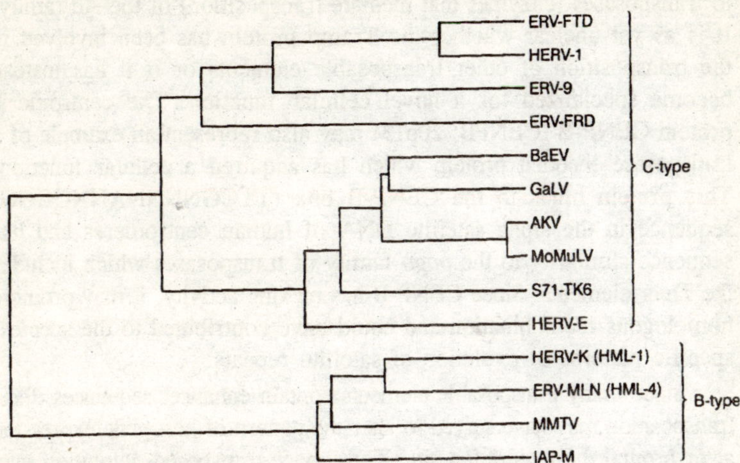

Fig. 5.4. Phylogenetic analysis of mammalian endogenous retroviral pol sequences.

mutational events. Another (not incompatible) view is that retroviruses themselves originally arose from intracellular retroelements (the protovirus hypothesis), a view which is supported by the phylogenetic analysis of endogenous retroviral DNA sequences. Regardless of whether or not the horizontal transmission of retroviral elements has taken place, copy number amplification has certainly occurred.

Retrotransposition of generates retroposon sequence variants owing to its inherent imprecision: target site rearrangements combine with the infidelity of both reverse transcriptase and RNA-dependent RNA polymerases to ensure that the inserted sequences are highly variable thus providing new avenues for the highly opportunistic evolutionary process.

Transposons

Transposon-like THE-1 repeats, which lack any obvious homology to retroviral sequenoes, have been found in the deletion-prone intron 43 of the dystrophin (DMD; Xp21.2-p21.3) gene, the human blood group GC (4q12) gene and the 3' untranslated region of the human calmodulin-protein (CALML1; 7p13-pter) gene. A cluster of three THE-1 repeats located in a 26 kb region of intron 7 in the human DMD gene has arisen by three independent insertion events. There is some evidence to support the hypothesis that the insertion of these elements has occurred at preferred target sites.

A pseudoautosomal gene sequence (*Tramp*) has recently been isolated which encodes within its single exon a protein with homology to transposases (enzymes that mediate transposition) of the Ac family. It is as yet unclear whether the Tramp protein has been involved in the transposition of other transposable elements or if it has instead become specialized for a novel cellular function. The centromeric protein CENP-B (CENPB; 20p13) may also represent an example of a transposase-encoded protein which has acquired a cellular function. This protein binds to the CENP-B box (TTCGNNNNANNCGGG) sequence in the alpha satellite DNA of human centromeres and has sequence similarity to the *pogo* family of transposases which includes the *Tiger* elements. Since CENP-B has nicking activity, it may promote homologous recombination and could have contributed to the species-specific patterns of evolution of satellite repeats.

Since many transposable elements contain enhancer sequences, their transposition may have served to alter the pattern of host gene expression at or around the integration site. Thus, once transposed, evolution may have recruited such enhancers to play a role in the transcriptional regulation of a gene in the vicinity of the integration site. One example of this is the human salivary amylase (AMY1C; 1p21) gene where the HERV-E-derived enhancer may be involved in tissue-specific expression.

Evolution may also recruit transposable elements as a means to alter mRNA processing. For example, a B2 (SINE) element has become inserted into the 3' untranslated region of the murine (*lifr*) gene encoding the soluble form of the leukemia inhibitory factor receptor (LIFR). Insertion of the B2 element has, by potentiating alternative 3' mRNA processing and alternative splicing, given rise to a truncated mRNA species (relative to the mRNA encoding the membrane-anchored LIFR) which encodes soluble LIFR. In the rat, no such retrotranspositional event has occurred and the soluble form of LIFR is not found.

A very special case of the opportunistic recruitment of a transposable element may have been that of the recombination-activating gene (RAG) transposase postulated to have been inserted into an ancestral immunoglobulin/T-cell receptor gene soon after the divergence of jawed and jawless fishes. The subsequent conversion of this transposon into a site-specific recombinase may have been the critical event in allowing the vertebrates to generate the genetic diversity so essential for the flexible adaptive response of their immune systems.

LINE Elements

LINE elements have assumed considerable importance in the context both of gene pathology and gene evolution. They are present in a wide range of mammals and are represented by some 40 subfamilies. They are nonrandomly distributed in the human genome, inserting preferentially into chromosomal G bands and at the DNA level, into A-rich sequences. The total numbers of LINE elements in four of the great apes have been estimated by Hwu et al. (1986): human, 107,000; chimpanzee, 51,000; gorilla, 64,000; and orangutan, 84,000. Since these figures differ markedly, it follows that numerous insertions and deletions of these sequences must have occurred during the evolution of the great apes.

In a pathological context, a number of examples of gene inactivation through insertion of LINE elements into a gene coding sequences are known and in some cases a preference for integration at AT-rich sequences is exhibited. Further, the target sites of two LINE elements inserted into the factor VIII (F8C; Xq28) gene causing hemophilia A are 80% homologous to a 10 nucleotide motif (GAAGACATAC) present in one of the highly favoured retroviral insertion target sequences reported by Smith et al. (1988).

Preferential target sites for LINE elements are also apparent in mammalian genes during evolution. For example, the interleukin-6 genes of rodents represent hotspots for LINE element retrotransposition. During mammalian evolution, the introduction of LINE elements in the vicinity of genes has sometimes altered gene expression as a consequence of their being recruited to perform a regulatory function. LINE elements have served to promote genetic rearrangements and indeed they may well have mediated both gene inversion and duplication events during evolution. Finally, and perhaps most importantly, LINE elements may have repeatedly transduced exons from one genomic location to another, thereby potentiating exon shuffling, the transfer of exons encoding specific protein modules between genes.

Alu Sequences

Evolution of Alu sequences

The fossil *Alu* monomer is thought to have arisen by the deletion of the central S domain of 7SL RNA (RN7SL) followed by the addition of a 3' poly(A) tract which may have facilitated reverse transcription of these RNA polymerase III transcripts. The free left arm monomer then arose by deletion of 42 bp from the fossil *Alu* monomer whilst the free right arm monomer arose by deletion of 11 bp from the fossil *Alu* monomer. The first *Alu* sequence may have been formed by dimerization of a free left arm monomer with a free right arm

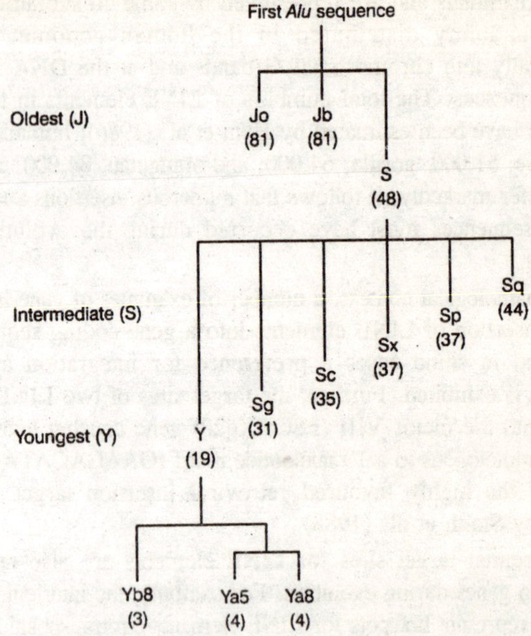

Fig. 5.5. The proposed evolution of the 12 human Alu subfamilies.

monomer an event which is thought to have occurred about 60 Myrs ago, before the divergence of prosimians. Subsequently, many rounds of sequential amplification took place to generate the 12 human *Alu* subfamilies seen today.

The total numbers of copies of Alu sequences in four of the great apes have been estimated by Hwu et al. (1986): human, 910,000; chimpanzee, 330,000; gorilla, 410,000; and orangutan, 580,000. As with the LINE elements, it would appear that numerous insertions and deletions of these sequences have occurred during the evolution of the great apes. At the chromosomal level, *Alu* sequences insert preferentially into R bands whereas at the DNA level, they preferentially integrate into A-rich sequences. During mammalian evolution, the introduction of *Alu* sequences in the vicinity of genes has sometimes altered gene expression as a consequence of their being recruited to perform a regulatory function. *Alu* sequences may also have been involved in, or mediated, many other different types of gene rearrangement during gene evolution including gross deletions, duplications, transpositions, gene fusions, recombination, and gene conversion events.

Alu sequence polymorphisms

Once inserted, specific *Alu* sequences have often been relatively stable in terms of their location during primate evolution. This notwithstanding, some human *Alu* sequences are polymorphic in terms of their presence or absence, a situation which is sometimes also found in other primates. Some of these polymorphisms may be of functional significance e.g. a common insertion/deletion polymorphism (0.4/0.60) within intron 16 of the human angiotensin I converting enzyme (DCP1; 17q23) gene, explicable in terms of the presence or absence of a 287 bp *Alu* repeat, is known to have an important influence on serum enzyme concentration. *Alu* sequence retrotransposition is also an occasional cause of genetic disease.

Alu sequence target sites

Although *Alu* sequences occur on average every 3-6 kb, there are several examples of regions that appear to be preferential target sites for *Alu* sequence insertion in mammalian genes. Thus, a 40 kb region, spanning the spermatid-specific protamine genes PRM1, PRM2 and the transition protein (TNP2) gene (16p13.2) contains a total of 42 *Alu* sequences. Similarly, a 22 kb region telomeric to the HLA-B associated transcript 2 (BAT2; D6S51E; 6p21.3) gene in the HLA class III locus contains 42 *Alu* repeats, whilst a 2.2 kb segment 5' to the human lysozyme (LYZ; 12) gene contains four such repeats.

Alu sequences within protein-coding sequences

Many *Alu* sequences are found within introns and therefore this repeat is represented in heterogeneous nuclear RNA. *Alu* sequences are however also found at different locations within mRNA-homologous sequences, the majority occurring within the untranslated regions (UTRs). Thus, Yulug et al. (1995) reported that 5% of full-length human cDNAs contained an *Alu* sequence, with 82% and 14% of these being located in the 3' UTR and 5' UTR, respectively. In a few cases, however, *Alu* sequences have been incorporated into the coding sequences of human genes and have therefore altered the amino acid sequences of the encoded proteins. Thus, a 279 bp *Alu* sequence spans 103 bp of the coding region of a zinc finger protein (ZNF91; 19p12) gene and extends 166 bp into the 3' UTR. Similarly, 110 bp of *Alu* sequence lies within the coding a region of the lectin-like type II integral membrane protein (KLRC1; 12) gene with 43 bp extending into the 3' UTR. An *Alu* sequence is entirely contained within the coding region of the protein serine/threonine kinase stk2 (STK2; 3p21.1) (279 bp *Alu*) gene. Finally, and perhaps most dramatically, two *Alu* sequences (both with poly(A) tails) are entirely contained within the coding region of the regulator of mitotic spindle assembly 1 (RMSA1; 17p11.2-p12) gene accounting for 111 amino acids of its coding potential, some 40% of the total.

Splice-mediated insertion of Alu sequences

Alu sequences have also been found to alter protein coding sequences through the ***splice-mediated insertion*** of the repeat and this probably represents the major mechanism by which *Alu* sequences have entered protein coding regions. The principle involved by reference to the pathological example of ornithine δ-aminotransferase deficiency caused by a single base-pair substitution in an intronic *Alu* element in the human ornithine δ-aminotransferase (OAT; 10q26) gene; this lesion activates a cryptic donor splice site which results in the incorporation of the *Alu* sequence into the mRNA.

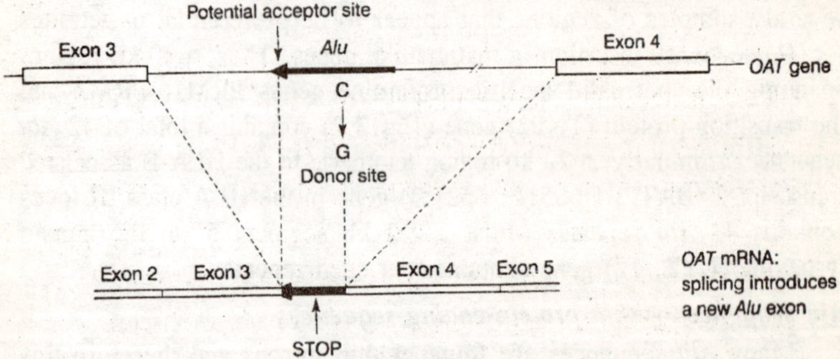

Fig. 5.6. Ornithine δ-aminotransferase deficiency caused by a mutation in a resident intronic Alu element within the OAT gene.

In an evolutionary context, there are several examples of the splice-mediated insertion of *Alu* repeats into human gene coding sequences. The splice-mediated insertion of a 95 bp *Alu* sequence has been reported in the lecithin: cholesterol acyltransferase (LCAT; 16q22.1) gene. In humans, the alternate *Alu*-containing transcript represents between 5% and 20% of the LCAT mRNA. It is also present in LCAT mRNA from chimpanzee, gorilla and orangutan; in the latter, the *Alu*-containing mRNA species constitutes 50% of the total LCAT mRNA pool. It into however present in the LCAT genes of gibbons, or Old World and New World monkeys.

In the human biliary glycoprotein (BGP; 19q 13.2) gene, three mRNA variants are produced as a result of the alternative splicing of an exon (IIa) with one of two virtually identical *Alu* cassettes derived from two intronic repeats. Other such examples are to be found in the

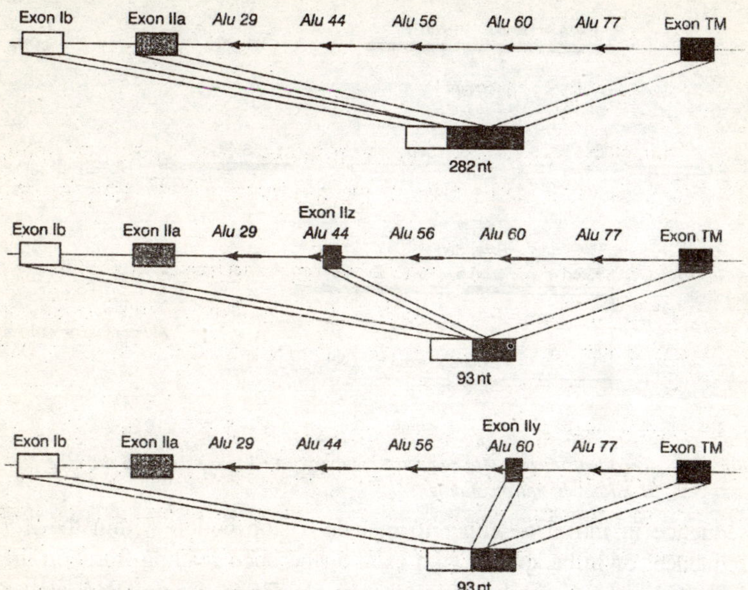

Fig. 5.7. A scheme for alternative splicing in the human biliary glycoprotein (BGP) mRNA.

human REL (2p12-p13) proto-oncogene, and the complement decay-accelerating factor (DAF; 1q32) and complement C5 (C5; 9q33) genes. Of the 17 *Alu* sequences found in mRNA coding regions by Makalowski et al. (1994), seven contained in-frame Stop codons and three others were predicted to cause frameshifts. Thus it is perhaps not surprising that in several cases of mRNA containing *Alu* sequences, *allelic exclusion* is evident and the mRNA containing the *Alu* sequence is of low abundance compared to splice variants of the same gene that lack the repeat. This notwithstanding, it may well be that the splice-mediated insertion of *Alu* repeats has been an important evolutionary mechanism for creating diversity at the protein level.

***Alu* sequence incorporation by intron sliding**

An alternative mechanism of *Alu* sequence incorporation into gene coding regions is intron sliding, and is illustrated by the example of the human HLA-DRB1 (6p21.3) gene; an intronic *Alu* sequence has been incorporated into exon 4 of the HLA-DR-β1 mRNA. Among three variants of the HLA-DR-β1 cDNA, detected by library screening, one was considered to be the usual form, whereas two others were alternatively spliced owing to a lack of splicing at the intron 5 donor site. As a result, exon 5 was extended into a nearby downstream *Alu*

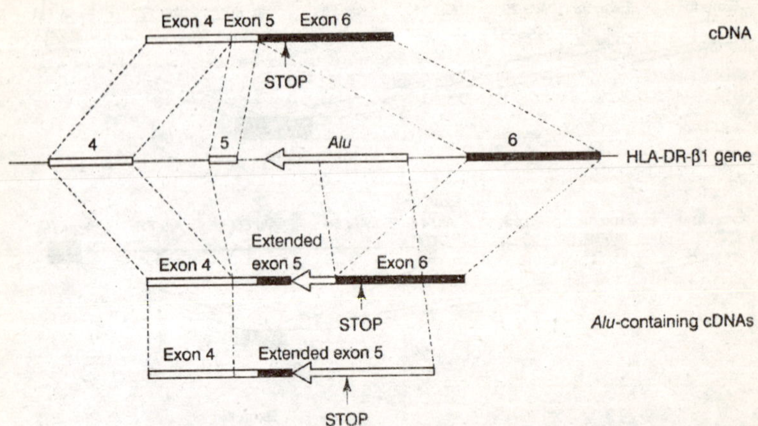

Fig. 5.8. A resident intronic Alu sequence is incorporated into exon 4 of the HLA-DR-β1 mRNA by intron sliding.

sequence in intron 5, either to include a stop codon within the *Alu* sequence, or to be spliced with exon 6 (the open reading frame in the extended exon 5 matches that of exon 6). These three cDNA clones may thus illustrate two phases of intron sliding: the inactivation of an existing splice site followed by the activation of a cryptic one.

Gross Gene Duplications in Evolution

Gene duplication (or partial duplication) events are a fairly uncommon cause of human genetic disease. Two distinct mechanisms are currently envisaged: (i) *homologous unequal recombination* either between homologous chromosome or sister chromatids and (ii) *nonhomologous recombination* at sites with minimal homology. Topoisomerase cleavage sites have been reported to be associated with pathological gene duplications and potential sites for topoisomerases I and /or II have been found to coincide with the breakpoints of duplication in the human factor VIII (F8C; Xq28) and dystrophin (DMD; Xp21.2-p21.3) genes. These observations are potentially interesting since topoisomerase activity has been implicated in several cases of nonhomologous recombination.

One of the best studied gross duplications in human genome pathology is the 1.5 Mb duplication of the short arm of chromosome 17 associated with Charcot-Marie-Tooth disease type 1A. This recurring duplication is thought to be mediated by homologous unequal recombination between two misaligned ~30 kb CMT1A-REP repeat sequences flanking the CMT1A region in direct tandem orientation. In humans, these repeats are 98% identical. Chimpanzees have two copies of a CMT1A-REP-like sequence, whereas gorilla, orangutan and gibbon only have a single copy consistent with a duplication of the CMT1A-REP sequence after gorilla diverged from the human lineage but before the divergence of chimpanzee and human. Orthologous sequence comparison has provided evidence that the distal repeat was the progenitor copy.

Duplications and the Emergence of Paralogous Genes

During vertebrae evolution, novel genes have arisen by genome duplication intra-chromosomal region duplication, and localized individual gene duplication. All three mechanisms give rise to *paralogous* genes, genes that occur within the same species and which have a common ancestor. Paralogous genes therefore include the members of multigene families and superfamilies. Evidence for the common ancestry of paralogous genes may come from sequence homologies (e.g. as with the voltage-sensitive ion channel genes) and/or from similar exon-intron organization, for example the cholesterol ester transfer protein (CETP; 16q21) and the phospholipid transfer protein (PLTP; 20q12-q13) genes, or the growth hormone receptor (GHR; 5p12-p14), prolactin receptor (PRLR; 5913-p14) and interferon receptor α, β, and ω, 1 (IFNAR1; 21q22.1) genes.

Intra-chromosomal Regional Duplication

In the human genome, whole chromosomal segments have sometimes been duplicated resulting in a series of paralogous genes retaining their syntenic arrangement. Thus, a number of genes located at 6p21.3 have paralogous genes at 9q33-q34; the chromosome 6 loci include genes for type 11 collagen α2 subunit, (COL11A2), NOTCH4, 70 kDa heat shock proteins (HSPA1A, HSPA1B, HSPA1L), valyl-tRNA synthetase 2 (VARS2), complement components (C2, C4A, C4B), pre B cell leukemia transcription factor 2 (PBX2) and retinoid X receptor β (RXRB) whilst the chromosome 9 paralogues include COL5A1, NOTCH1, HSPA5, VARS1, C5, PBX3, and RXRA. Other extensive chromosomal duplications included genomic segments present at Xq28 and 16p 11.1 involving the paralogous creatine transporter genes SLC6A8 and SLC6A10 respectively. Some intra-chromosomal duplications may only involve one or a small number of genes, for example the duplication of the iduronate-2-sulphatase (IDS) locus at Xq28. Another example is that of the inverted duplication at 5q13 which duplicated the spinal muscular atrophy (SMA) gene, the survival motor neuron (SMN) gene and the apoptosis inhibitory protein (NAIP) gene. Duplicated paralogous genes may however be translocated to quite different locations on the same chromosome, for example the adrenergic receptor (ADRA1B and ADRB2) genes on 5q23-q32 are quite distant from the evolutionarily related serotonin receptor (HTR1A) gene on 5cen-q11.

Tandem Duplications

Multigene families often form syntenic gene clusters as a result of the tandem duplication of an ancestral gene sequence. For example, the immunoglobulin genes are clustered at 14q32.33 (IGHA, IGHD, IGHG) and 22q11.2 (IGLC, IGLL), the T-cell receptor genes at 14q11.2 (TCRA, TCRD), 7q35 (TCRB) and 7p14-p15 (TCRG), the pregnancy-specific glycoprotein (PSG1, PSG2, PSG3, PSG4, PSG5, PSG6, PSG7, PSG8, PSG11, PSG12, PSG13) genes at chromosome 19q13.2, the histocompatibility antigen (HLA) genes at chromosome 6p21.3 whilst the three alkaline phosphatase genes (ALPP, ALPI, ALPPL2) are clustered at chromosome 2q37.

Some genes appear especially prone to duplicate, probably by virtue of their already being clustered in multiple copies. Thus, the carcinoembryonic antigen (PSG, CEA; 19q13.2) gene family has undergone multiple, but independent, multiplication events in both the rodent and primate lineages. In similar vein, a disproportionate fraction of mapped zinc finger gene family members are located on chromosome 19. Such clustering has probably arisen as a result of the serial duplication of a single ancestral gene on the same chromosome.

Members of gene families in different species may however be amplified differentially. For example, in the human genome, three members of the formylpeptide receptor gene family (FPR1, FPRL1, and FPRL2) cluster at chromosome 19q13.3, whereas in mouse, there are six *Fpr* gene on a syntenic region of chromosome 17. The human FPRL2 gene and four murine *Fpr* genes arose after the divergence of human and mouse.

Translocation of Duplicated Genes

Gene family members have often become chromosomally separated through translocation and it would appear as if the older the multi gene family, the more likely this is to happen. Thus, the chromosomally unlinked CD36L2 (chromosome 4), CD36L1 (chromosome 12) and CD36 (7q11.2) membrane glycoprotein genes were duplicated and diverged from an ancestral gene prior to the separation of the arthropod and vertebrate lineages. Other examples of ancient duplicated genes that have become separated by translocation are the thrombospondin

(THBS1, 15q15 and THBS2, 6q27) genes which arose ~900 Myrs ago and the transforming growth factor-β (TGFB1, 19q13.2; TGFB2, 1q41; TGFB3, 14q24) genes which arose ~30 Myrs ago.

Duplication of Translocated Genes

Genomic duplication may be followed by further local tandem duplication as exemplified by the human homeobox gene cluster 7p14-p15 (HOXA), 17q21-q22 (HOXB), 12q12-q13 (HOXC) and 2q31 (HOXD). The same phenomenon is exhibited by the human sodium channel genes many of which reside in the same paralogous chromosome segments as the HOX gene clusters: SCN1A, SCN2A, SCN3A, SCN6A, SCN7A and SCN9A (2q23-q24), SCN8A (12q13), SCN4A (17q23-q25), SCN5A and SCN10A (3p21-p24).

By contrast the localized duplication of translocated genes appears to be a general features of the human olfactory receptor (OLFR) genes. Thus, several OLFR gene clusters on chromosome 11 (11p15, 11p13, 11q24) are more similar to each other than to OLFR genes on chromosome 17, implying that translocation was followed by regional tandem duplication. However, the OLFR gene cluster on chromosome 17p13.3 contains members that do not belong to the same subfamily, suggesting that it originated instead by duplication of an entire gene cluster followed by translocation of that cluster.

Syntenic Relationships and Gene Dispersal

Gene duplication sometimes creates gene families whose members are both syntenic and dispersed, for example the human purinoceptor gene family, (P2RY1, 3; P2RY2, 11q13.5-q14.1; P2RY4, Xq13; P2RY6, 11q13.5, P2RY17, chromosome 14). Seven human matrix metalloproteinase genes cluster at chromosome 11q22.3 (MMP1, MMP3, MMP7, MMP8, MMP10, MMP12, MMP13), but the other family members are dispersed between many other chromosomes (MMP2, 16q13; MMP9, 20q11.2-q13.1; MMP11, 22q11.2; MMP14, 14q11-q12; MMP15, 16q21; MMP16, 8q21; MMP19, 12q14). Other examples include the human fucosyltransferase (FUT1, FUT2, FUT3, FUT5, FUT6, 19p13.3; FUT4, 11q21; FUT7, 9q34; FUT8, 14q23) and annexin genes (ANX1, 9q11-q22; ANX2, 15q21-q22; ANX3, 4q13-q22; ANX4, 2p13; ANX5, 4q28-q32; ANX6, 5q32-q34; ANX7, 10q21.1-q21.2; ANX8, 10q11.2; ANX11, 10q21.1-q21.2; ANX13, 8q24.1-q24.2).

In some cases, evolutionary conservation of post-duplicational clustering may be important for the coordinate regulation of individual genes by common control elements. Possible examples of this include the spermatid-specific protamine (PRM1 and PRM2; 16p13.2) genes, the platelet membrane glycoprotein (ITGA2B, and ITGA3; 17q21-q22) genes, the albumin gene family (ALB, AFP, AFM, GC; 4q11-q13), the pregnancy-specific glycoprotein (PSG1, PSG2, PSG3, PSG 4, PSG5, PSG6, PSG7, PSG8, PSG11, PSG12, PSG13; 19q13.2) genes and the fibrinogen α, β, and γ (FGA, FGB and FGG; 4q21) genes.

Individual duplicated genes may still exhibit synteny simply because recombination has not yet separated them. Thus, the paralogous pulmonary surfactant protein D (SFTPD) gene at 10q22-23 lies in very close proximity to the pulmonary surfactant protein A (SFTPA1) gene. Similarly, the paralogous the interferon α/β receptor gene (IFNAR1) is closely linked to the cytokine receptor B4 (CRFB4) gene on chromosome 21q22.1. Synteny may however be retained for very long periods of evolutionary time. Indeed, the human proteasome $\alpha 2$ subunit (PSMA2) and TATA box-binding protein (TBP) genes are linked on chromosome 6q27 and their orthologues are also syntenic in *Drosophila melanogaster* and *Caenorhabditis elegans*. Similarly, conserved synteny is also apparent between the human genes PIM1, RXRB, PBX2, NOTCH4 and TNXA at 6p21 and their orthologues in *D. melanogaster* and *C. elegans*. The evolutionary conservation of close linkage over such long time periods is highly unusual and implies the existence of underlying functional reasons.

Functional Redundancy and Post-duplication Diversification

Evolution is opportunistic and gene duplication provides the opportunity for structural and functional diversification. This is well exemplified by the origin of the α-lactalbumin gene (LALBA; 12q13). α-Lactalbumin may be regarded as essentially a mammalian ‘invention’. It is a regulatory subunit of the enzyme lactose synthetase. Phylogenetic analysis has indicated that α-lactalbumin evolved from lysozyme (LYZ; chromosome 12) (a bacteriolytic enzyme present in both vertebrates and invertebrates) through a gene duplication event that occurred before the divergence of mammals and birds, and prior to the evolution of the mammary gland. The origin of the LALBA gene thus appears to have long preceded the acquisition of its modern function. Its rapid evolution in the mammalian lineage need not however have been solely due to its acquisition of a new function. Once its lysozyme activity was lost, many of the amino acids essential for the protein’s original function were freed from selective pressure, allowing rapid change, not all of it necessarily adaptive.

The protein products of a gene duplication may sometime be targeted to different cellular locations. The human mitochondrial HMG

CoA synthetase (HMGCS2; 1p12-1p13) gene encodes an enzyme which catalyses the first reaction of ketogenesis whilst the cytoplasmic form of the enzyme, encoded by the HMGCS1 gene (5p13), catalyses the second step in cholesterol biosynthesis from acetyl CoA. The two genes are thought to have arisen from a gene duplication ~500 Myrs ago. The emergence of the two enzymes as distinct entities served to link the pathways of β-oxidation and leucine catabolism and created the HMG CoA pathway of ketogenesis thereby providing a lipid-derived energy source.

Gene duplication has also potentiated the emergence of *isozymes*, enzymes that catalyze the same biochemical reaction but which may differ from each other in terms of tissue specificity, developmental regulation or biochemical properties. Thus, in vertebrates, the two subunits of lactate dehydrogenase are encoded by two genes (LDHA; 11p14-p15 and LDHB; 12p12) and these subunits can be combined in such a way as to form five tetrameric isozymes (A_4, A_3B, A_2B_2, AB_3, and B_4) each with its own distinctive properties.

Iwabe et al. (1996) examined gene duplications during organismal evolution and concluded that most gene duplications giving rise to entirely novel functions predated the divergence of the vertebrate and arthropod lineages. Genes which encode protein that are localized to cell compartments (compartmentalized isoforms) emerged by duplications which predated the separation of animals and fungi. By contrast, genes encoding products with virtually identical functions but differing tissue distribution (tissue-specific isoforms) have undergone duplications independently in vertebrates and arthropods after divergence of the vertebrate and arthropod lineages. Iwabe et al. (1996) concluded that there was a good correspondence between molecular evolution at the level of the gene, and tissue and organismal evolution.

Several gene duplications have occurred during primate evolution. For example, the tandem duplication responsible for the creation of the γ1-(HBG1) and γ2-(HBG2) globin genes occurred prior to the divergence of Old World from New World monkeys. The 5.5 kb duplicated segment is bounded by two related LINE elements suggesting that the duplication occurred via homologous unequal recombination. Perhaps it was the functional redundancy initially introduced by the duplication which created an opportunity for the γ-globin genes to evolve a fetal function to replace their original embryonic function. Post-duplicational redundancy is sometimes still apparent in some systems. For instance, the MyoD family of transcription factors involved in myogenesis in skeletal muscle still exhibits functional redundancy (between the proteins encoded by the MYOD1 (11p15.1), MYF5 (chromosome 12) and MYF6 (chromosome 12) genes) and this is probably indicative of overlapping functions.

The primate glycophorin genes (GYPA, GYPB, GYPE; 4q28-q31) are thought to have emerged by duplication, mediated perhaps by recombination between *Alu* sequences. GYPA probably represents the ancestral gene and is present in all primates studied. The GYPB gene is present in human, chimpanzee and gorilla but not orangutan and gibbon, whereas the GYPE gene is present in human, chimpanzee but intriguingly only 7/16 gorillas tested. The complement C4 (C4A, C4B; 6p21.3) genes and the cytochrome CYP21 (6p21.3) gene also emerged in the primate lineage as a result of a single duplication occurring prior to the divergence of the apes from the Old World monkeys.

There are several other examples of gene duplications occurring in only one mammalian order. For example the bovine-specific coglutinin (*Cgn1*) gene is homologous to the human pulmonary surfactant protein D (SFTPD; 10q22-q23) gene and is thought to have evolved by gene duplication in the Bovidae after their divergence from the other mammals. Sometimes the number and distribution of *Alu* sequences may help the reconstruction of the phylogeny of a gene duplication event as in the case of the duplication of the apolipoprotein CI (APOC1; 19q13.2) gene, timed at about 39 Myrs ago. Some partially homologous genes may have undergone a process of duplication and divergence but one or other copy may have either acquired or lost DNA sequence at some stage thereby limiting the extent of observable homology between them. One example of this is the human α-N-acetylgalactosaminidase (NAGA; 22qter) and α-galactosidase A (GLA; Xq) genes. Six of the seven GLA exons were identically positioned in the NAGA gene but there was no similarity between the predicted amino acid sequences of GLA exon 7 and NAGA exons 8 and 9.

Truncated Gene Copies

In some cases, only a portion of a gene may be involved in these intra-chromosomal duplication event [e.g. the polycystic kidney disease 1 (PKD1; 16p13.3) gene)]. This often results, as in the PKD1 example cited, in the generation of a linked pseudogene representing a partial copy of the parental source gene . However, truncated copies need not always be inactive. Thus, the melanin-concentrating hormone (PMCH; 12q23-q24) gene has become partially duplicated during primate evolution to generate two truncated copies (PMCHL1 and PMCHL2) which have been translocated to chromosome 5p14 and 5q12-q13 respectively. These variant gene copies possess open reading frames, are expressed in a distinctive tissue-specific fashion and, in the authors' opinion, may represent 'genes in search of a function'.

Duplication Polymorphisms

Some gene duplications occur as polymorphic variants in the human population, for example the red (RCP; Xq28) and green (GCP; Xq28) visual pigment gene. Human males with trichromatic vision typically possess one RCP gene, one, two or more GCP genes and an RCP/GCP hybrid gene. However, some males with apparently normal colour vision can possess 4 RCP genes and 6 or 7 GCP genes. This notwithstanding, only one GCP gene is normally expressed, probably as a result of the activity of a locus control region upstream of the RCP gene.

Other examples of duplicational polymorphisms include the α1-globin, (HBA1; 16p13.3;), ζ-globin (HBZ, 16p13.3;) Gγ-globin (HBG2; 11p15), haptoglobin (HP; 16q22.1) and proline-rich protein (PRB1, PRB2, PRB3, PRB4; 12p13.2) genes. Duplicational polymorphisms may even manifest as gene cluster copy number variation, for example that involving the CYP21, CYP21P, C4A, and C4B genes at human chromosome 6p21.3. Another example of a duplicational polymorphism is that of three members (one potentially functional) of the olfactory receptor (OLFR) gene family which occur in tandem within a block that is duplicated at 14 different subtelomeric location in the human genome. This results in normal individuals possessing between 7 and 11 copies of this block in their genomes. Trask et al. (1998a) suggested that sub-telomeric regions could serve as 'nurseries' for the generation of diversity by promoting gene duplication.

Other gross duplicational polymorphisms are found in the immunoglobulin V_H gene cluster (IGHV; 14q32): one of 50 kb in length is present in 73% of individuals and results in the gain of 5 functional V_H segments whilst another of length ~80 kb is present ~50% of individuals and involves the gain/loss of two functional V_H segments. A third such polymorphism involving only a single additional V_H segment has been reported to occur in 27% of individuals. A duplicational polymorphism is also apparent in the C_H gene cluster: the IGHG4 gene (14q32) is duplicated in 44% of haplotypes. Finally, the partial duplication of the GABA A receptor α5 (GABRA5) gene is polymorphic in that individuals differ with respect to gene copy number.

INTRAGENIC GENE DUPLICATIONS IN EVOLUTION

Numerous examples of human genes have now come to light in which the encoded proteins have emerged through the introduction of individual exons or blocks of exons. Some of these exons encode specific protein domains (e.g. zinc finger, homeobox, immunoglobulin-like, epidermal growth factor-like, fibronectin, ABC cassette, Sushi, ankyrin, chymotrypsin etc.) which have come to be distributed between a large number of different protein through *exon shuffling*. Protein evolution may however also involve the internal duplication or amplification of individual eons, blocks of exons or alternatively repetitive sequence motifs within exons. Some typical examples of such intragenic duplication event are explored below.

Multi-exon Duplications

Some genes encode proteins that comprise two homologous domains and are therefore likely to have originated through a gross internal duplication. Thus, the 17 exon human transferrin (TF; 3q21) gene is thought to have originated by an internal duplication which may have resulted from an unequal crossing over event. Similarly, the human angiotensin I converting enzyme is encoded by a gene (DCP1; 17q23) which comprises 26 exons that encode two homologous domains each containing an active site. Exon number and size in the DCP1 gene are consistent with an ancient internal duplication, as are the codon phases at the exon-intron boundaries. An ancient internal duplication probably also occurred in the cystic fibrosis transmembrane conductance regulator (CFTR; 7q31.3) gene which encodes a protein with two transmembrane domains and two nucleotide binding fold domains. By contrast, successive duplications of ancestral domains appear to have occurred

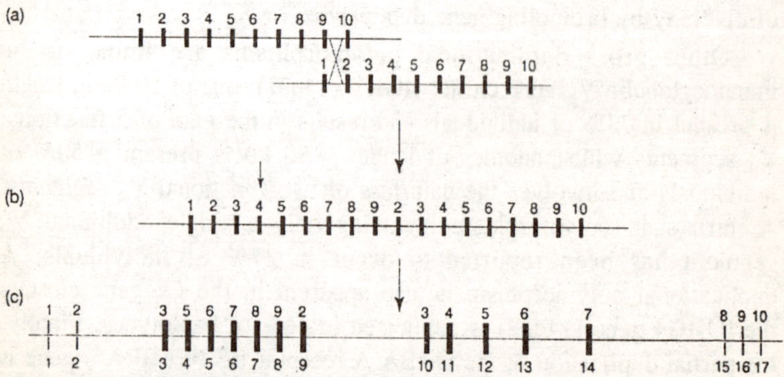

Fig. 5.9. A possible scheme for the evolution of the human transferrin (TF) gene.

in the human kininogen (KNG; 3q27) gene. Other examples of human proteins displaying internal domain duplication include calbindin (six 43 amino acid repeats), fibronectin (twelve 40 amino acid repeats), plasminogen (five 79 amino acid repeats) and α-tropomyosin (seven 42 amino acid repeats). Many other genes are also likely to have evolved by internal gene duplication but the duplicated regions have probably diverged so much over evolutionary time that sequence homology between them is no longer discernible.

Exon Duplication

It has been estimated that at least 6% of exons in human genes have arisen by the duplication of pre-existing exons. One example is that of the human CHC1 (1p36.1) gene, which encodes a protein involved in the coupling between DNA replication and mitosis. This gene comprises 14 exons, eight of which encode the 7 tandemly repeated domains of ~60 amino acids within the CHC1 protein; each repeat is encoded by a single exon except for repeat IV which is encoded by exons 10 and 11 separated by an inserted intron. The CHC1 gene

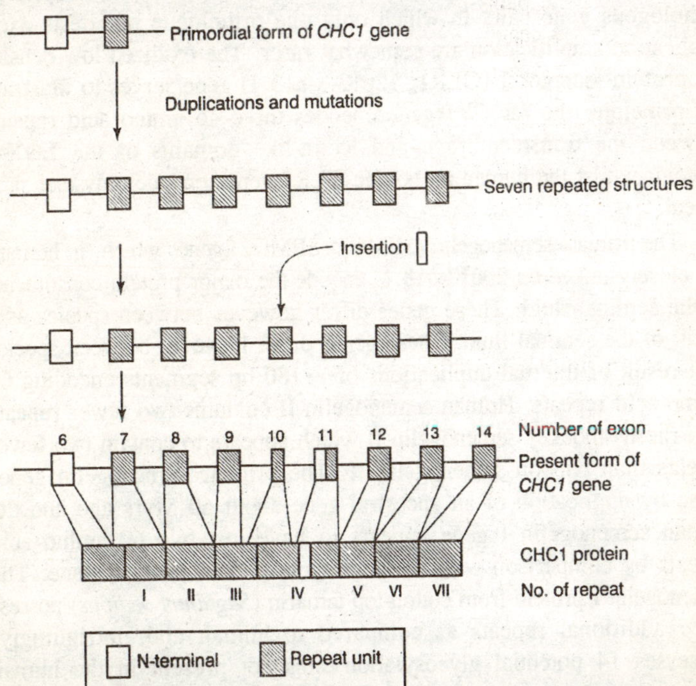

Fig. 5.10. Structure and evolution of the human CHC1 gene.

therefore appears to have arisen through the amplification of a primordial exon. Gene construction by exon amplification has also been employed in the macrophage mannose receptor (MRC1; 10p13) gene; 26 of the 30 exons of the MRC1 gene serve to encode the eight C-type carbohydrate recognition domains. Ceruloplasmin (CP; 3q23-q25) consists of three homologous repeat units and appears to have evolved by successive duplication of a primordial copper-binding protein domain of ~350 amino acid residues. Finally, the human annexin II (ANX2; 15q21-q22) gene encodes a protein containing 8 copies of a conserved 70 amino acid repeat.

Exon duplication/ amplification is clearly a common mechanism of evolutionary change and its occurrence may often be inferred from studies of exon/intron distribution or from the repetitivity of protein domains. Thus, the six paralogous genes of the human salivary proline-rich protein family closely linked on chromosome 12p13.2, differ from each other in terms of the number of copies of a 63 bp tandem repeat in exons 3 of the gene: PRH1 and PRH2 (6 repeats), PRB1 and PRB3 (15 repeats) PRB2 (16 repeats) and PRB4 (11 repeats). Examples of orthologous gene pairs in which only one orthologue manifests exon duplication/amplification are somewhat rarer. The oxidized low density lipoprotein receptor 1 (OLR1; 12p12.3-p13.1) gene serves to illustrate the principle; the rat *Olr1* gene encodes three 46 amino acid repeats between the transmembrane and lectin-like domains of the LOX-1 protein, whilst the human an bovine OLR1 genes encode only one such repeat.

The primate semenogelin (SEMG1, SEMG2) genes which, in human, are closely linked on 20q12-q13.1, encode the major protein constituents of the seminal fluid. These genes differ however between species as a result of the seminal fluid. These genes differ however between species as a result of internal duplications of ~180 bp segments encoding 60 amino acid repeats. Human semenogelin II contains two fewer repeats than rhesus monkey semenogelin II which appears to contain two fewer repeats than baboon semenogelin II. The primate semenogelin genes arose by duplication of an ancestral gene about 60 Myrs ago and the human semenogelin I gene appears to have lost two 60 amino acid repeats by comparison with the paralogous semenogelin II gene. The semenogelin I protein from cotton-top tamarin (*Saguinus oedipus*) possess three additional repeats as compared to human and, intriguingly, possesses 14 potential glycosylation sites not present in the human protein.

Most known genes manifesting internal duplication exhibit strong conservation of intron positions between the duplicated domains. One notable exception, however, is the rabbit phosphofructokinase gene. Sequence homologies between bacterial and rabbit phosphofructokinases, and between the amino and carboxy terminal ends of the rabbit enzyme are consistent with the origin of this gene being via a process of internal duplication and divergence. However, intron positions are not conserved even between the two halves of the rabbit gene.

Exon duplication may potentiate alternative splicing. Thus, the human ketohexokinase (KHK; 2p23) gene contains two very similar 135 bp exons (termed 3a and 3c) which are mutually exclusively spliced into KHK mRNA. Both the exon-intron structure and the pattern of alternative splicing are conserved between human, rat and mouse, consistent with the existence of two evolutionarily conserved KHK isoforms. This alternative splicing event is also tissue-specific since, in both rat and human, those tissues that express high levels of KHK incorporate exon 3c whereas other tissues incorporate exon 3a. Interestingly, a shift in splicing choice from exon 3a to 3c appears to occur during development between the human fetus and adult.

Intra-exonic Duplications

Internal gene duplications may sometimes be quite subtle. Thus, the 3' untranslated regions of the human and monkey cytochrome c oxidase subunit II (MTCO2; mitochondrial genome) genes have generated by duplication events involving a 14 bp region that occurred during primate evolution. The RNA derived from the MTCO2 gene has the potential to form stable stem-loop structures in a region immediately preceding the duplication site and these inverted repeat sequences may have played a role in promoting the duplicational events.

In some cases, homogenization of internal repeats has occurred as for example with the complement control protein repeats of the baboon and human complement receptor type 1 (CR1; 1q32) genes. This process has been termed *horizontal* or *concerted evolution* and the mechanism involved in likely to be either unequal crossing over or gene conversion. Whichever mechanism is responsible, the following example suggests that intragenic homogenization may be a less efficient process than intergenic homogenization. The human small proline-rich (SPRR) proteins, induced during keratinocyte differentiation, are encoded by two SPRR1 genes (A and B), approximately seven SPRR2 genes and a single SPRR3 gene, closely linked on chromosome 1q21-q22. The central segments of the encoded polypeptides are composed of tandemly repeated units of either eight or nine amino acids. Thus the consensus octamer PKVPEPCH is found 6 times in SPRR1A and SPRR1B, the nonamer PKCPEPCPP three times in SPRR2 and the octamer TKVPEPGC 14 times in SPRR3. This is consistent with a process of internal duplication that began after the divergence of the SPRR genes into three distinct subfamilies. It is evident that, during the evolution of the SPRR gene family, there has been a bias toward either intragenic or intergenic duplications. Thus in SPRR2, intergenic recombination has occurred more frequently than intragenic recombination since there are seven SPRR2 genes each with three repeats. By contrast, in SPRR3, no intergenic recombination has occurred but intragenic recombination has occurred frequently to generate 14 repeat copies. Interestingly the percentage of amino acid conservation (relative to the consensus for each type of gene) is significantly higher for the SPRR2 repeats than for either SPRR1 or SPRR3 suggesting that intergenic homogenization may be more effective than intragenic homogenization.

Concerted evolution is not however an obligatory property of internally repetitive proteins. Take for example the case of the α- and β-spectrins. Spectrin is a red blood cell cytoskeletal component which consists of a tetramer of two antiparallel ab spectrin dimers. The α-spectrin (SPTA1; 1q21) and β-spectrin (SPTB; 14q22-q23) genes evolved by duplication of a common ancestral gene which existed before the divergence of the vertebrate and arthropod lineages ~600 Myrs ago. The structure of both protein subunits is consistent with successive intragenic duplications. The α- and β-subunits consist of tandemly repeated segments of 106 amino acids of which 20 occur in α-spectrin, 17 in β-spectrin. Although the α-spectrin segments appear to have evolved in homogeneous fashion, β-spectrin segments exhibit considerable heterogeneity. One explanation for this difference is that some segments with specific functions may have evolved differently from others. Indeed, on the basis of the similar locations of the heterogeneous α- and β-spectrin segments, Muse et al. (1997) suggested that the α- and β-spectrins have co-evolved, and those segments that are intimately involved in subunit dimerization have been evolutionarily constrained by the structures of their binding partners. Muse et al. (1997) found no evidence for interrepeat exchanges and therefore concluded that neither gene conversion nor recombination had operated. At some stage, probably before the divergence of arthropods and vertebrates, concerted evolution may have operated but this initial phase probably ceased as the individual segments began to diverge at the DNA level.

Emergence of Primordial Genes by Oligomer Duplication

It is possible that primordial coding sequences emerged by a process of sequential duplication of base oligomers to yield 'genes' encoding polypeptides with significant periodicity. Ohno (1987) cites the example of a putative primordial rhodopsin gene, which gave rise to this ancient family of seven transmembrane domain-containing proteins including among others, bacterial rhodopsin and vertebrae retinal opsin, β_2-adrenergic receptor and muscarinic acetylcholine receptor. This primordial gene originally contained CCTGCTG, CCTGGCC and GCTGGCC heptameric repeats. Ohno (1987) showed that the gene encoding porcine muscarinic acetylcholine receptor still contains many of these oligomeric repeats. The original heptameric repeats appear to be more stringently conserved in those portions of the gene encoding the seven transmembrane domains whereas new repeat units comingle with old repeats in those portions that encode the extracellular and intra-cytoplasmic domains.

Intragenic Duplication Polymorphisms

Partial internal duplications of genes can also occur as polymorphic variants. A particularly dramatic example is provided by the human apolipoprotein(a) (LPA; 6q27) gene. This gene possesses at least 34 different alleles containing a variable number (between 12 and 51) of tandemly repeated kringle IV-encoding domains. Each allele gives rise to a different apolipoprotein(a) isoform thereby explaining the size polymorphism of the protein: 300-800 kDa. The LPA gene has been described as a 'plasminogen gene gone awry'; an apt description in that the closely linked and evolutionary related plasminogen (PLG; 6q27) gene encodes a protein with only 5 kringles. The LPA gene is thought to have arisen between 40 Myrs and 90 Myrs ago during the adaptive radiation of the mammals. At least ten kringle IV-encoding domains are present in rhesus macaque suggesting that kringle number may have expanded progressively during primate evolution. Although the LPA gene was originally thought to be confined exclusively to primates, it has also been found in hedgehogs in which it possesses 31 kringle-encoding domains. Since these are kringle III repeats, we must surmise that an apolipoprotein(a)-like gene arose independently from a plasminogen-like gene in this insectivore ($\sim$80 Myrs ago) and experienced a similar expansion of a subset of its kringle repeats to that found in humans—a remarkable example of *convergent evolution*. In humans, the LPA kringle repeat number polymorphism may not be without clinical significance because (i) there is an inverse relationship between apolipoprotein(a) isoform size and plasma apolipoprotein(a) levels and (ii) elevated levels of apolipoprotein(a) are associated with an increased risk of atherosclerosis and cardiovascular disease.

Several human mucin genes exhibit highly polymorphic tandemly repetitive regions, for example MUC1 (1q21), MUC2 (11p15) and MUC4 (3q29). Intragenic repeat copy number polymorphisms are also evident in the human proline-rich protein (PRH1; 12p13.2) gene where alleles vary in terms of the number of copies of a 63 bp repeat in exon 3 and the complement receptor type 1 (CR1; 1q32) gene where the two most common alleles differ in terms of the presence of a long homologous repeat of ~450 amino acids. Finally, in the human filaggrin (FLG; 1q21) gene, a length polymorphism of 10, 11, or 12 copies of a 972 bp repeat occurs within its polyprotein precursor which is subsequently cleaved into individual functional filaggrin molecules.

Coding Sequence Expansion and Contraction Resulting from the Introduction or Removal of Initiation and Termination codons

The mutation of initiation codons leading to the extension of the protein coding sequence of a gene is known to be a cause, albeit an infrequent one, of human genetic disease. Such mutations have however also occurred on a number of occasion during gene evolution. Thus, an ATG→GTC substitution occurred in the Met initiator codon of the ZNF80 (3q13.3) gene in the common ancestor of African green monkey and rhesus macaque which resulted in a change in the site of translational initiation to a location 20 codons amino terminal to the Met initiator codon used by humans, chimpanzees and gorillas. However, in African green monkey, the ZNF80 protein is only 213 amino acid residues in length (as compared to 273 residues in humans and the great apes and 293 residues in rhesus macaque) owing to truncation of the protein due to an additional GAG→TAG substitution introducing a novel termination codon.

Another example of mutation resulting in the specific-specific use of alternative initiation codons is provided by an ATG→ATA transition in the human hepatic peroxisomal L-alanine: glyoxylate aminotransferase 1 (AGXT; 2q36-q37) gene. This lesion removed the original initiation codon and an alternative downstream initiation codon is used in the translation of the human AGXT transcript. The rat *Agxt* orthologue encodes a protein which, by comparison with the human protein, possesses an extra 22 amino terminal amino acids specifying a leader sequence thought to contain a mitochondrial targeting signal that is absent in the human protein.

The mutational removal of a termination codon may also lead to the elongation of a protein product. One example of this is provided by the porcine P-glycoprotein genes whose coding regions may be seen to have been extended by such a mutation when compared to the human orthologous proteins (PGY1, PGY3; 7q21).

Minisatellites, Microsatellites and Telomeric Repeats

A detailed discussion of the evolution of the various families of satellite, minisatellite and microsatellite DNA would be somewhat tangential to the remit of this volume and would merely serve to recapitulate a previous volume in this series. However, a brief resume will be given to provide the necessary background to guide the interested reader to the relevant literature.

Minisatellite DNA Sequences

Minisatellite DNA sequences occur in all eukaryotes from yeast to human. Minisatellites frequently exhibit substantial allelic variability with respect to repeat number and allele length analysis has demonstrated germline mutation rates as high as 15% per gamete. Minisatellite mutation may involve intra-allelic rearrangements whose frequency, unlike inter-allelic rearrangements, is influenced by the size of the tandem array. Sequence similarities, manifested by a subset of minisatellites, to the Chi recombination promoting element of *E. coli* have led to the suggestion that this 'core sequence' might be recombinogenic and could serve to promoter unequal crossing over. However, analysis of flanking polymorphisms has not indicated the exchange of markers predicted for the products of unequal exchange between alleles. This notwithstanding, recombination hotspots sometimes co-localize with minisatellites which has led to the suggestion that minisatellite instability may be a by-product of meiotic recombination.

Monckton et al. (1994) have shown that minisatellite mutation can involve complex inter-allelic gene conversion events. These may exhibit polarity since the gain of a few repeat units appears to be confined to one end of the tandem repeat array. One alternative proposal is that minisatellite mutation may involve an array homogenization process which could operate by biased repair of intra-helical (slippage) or inter-helical (unequal sister chromatid exchange) heteroduplexes. Whatever the mechanism(s) underlining their allelic variability, the rapid evolution of minisatellites appears to have rendered them an important substrate for the opportunistic processes of molecular evolution. Indeed, several are known to have become recruited as gene regulatory elements.

Microsatellite DNA Sequences

Microsatellites typically mutate with a frequency of 10^{-3} to 10^{-4} although mutation rates vary between loci by several orders of magnitude. Microsatellite mutation events in the male germline outnumber those in the female germline 5- to 6-fold. Interestingly, alternative alleles at the same locus may differ dramatically in terms of their mutation rate. By contrast, microsatellite location is often conserved at orthologous positions in the genomes of different species compared allele length distributions for a considerable number of human microsatellites with their orthologues in chimpanzees, gorillas orangutans, baboons, and macaques and claimed a tendency for the loci to be longer in humans. Although some microsatellites are more variable in nonhuman primates than in humans, a tendency for the loci to be longer in humans would be consistent with the directionality of microsatellite evolution and compatible with evolution proceeding at different rates in different species. However, it could also be due to ascertainment bias in that it is precisely the longest, most mutable and therefore most informative of human microsatellites that have been selected as genetic markers. In principle, the large size of the human population could have compounded this effect since it could support a higher level of genetic diversity; in smaller populations, much variability is lost as alleles go to fixation. In practice, however, it would appear as if ascertainment bias cannot be the sole explanation for inter-specific differences in microsatellite repeat length. Thus, the tendency for microsatellite loci to be longer in humans may not simply be an experimental artefact.

Telomeric and Centromeric Repetitive DNA

Telomeric TTAGGG repeat number varies not only between human chromosomal arms but also between individuals. Indeed, there is some evidence in humans for the exchange of telomeric and subtelomeric repeats between nonhomologous chromosomal ends. Since satellite DNA sequences at the telomeric junctions of chimpanzees do not show any similarity to their counterparts in human or orangutan, it is likely that telomeres became reorganized relatively recently during primate evolution.

Alphoid satellite DNA is found as a tandem repeat in long chromosome-specific arrays at the centromers of all primate chromosomes. These arrays are highly variable within and between homologous chromosomes in the same species.

Expansion of Unstable Repeat Sequences

DNA sequences that are internally repetitive are particularly prone to misalignment during DNA synthesis. If such misalignment takes place, the nascent strand can slip back in multiples of the repeat unit and the newly synthesized DNA strand will be elongated by comparison with the parental strand. Since many coding sequences contain simple sequence repeats (for example, the genes encoding the 28S and 18S ribosomal RNAs (RNR1-5) and the TATA-binding protein TBP (TBP)), repeat expansion may have been involved in the evolution of these sequence . However, these simple repeats also have the potential to be involved in disease pathogenesis and it is from studies of genetic disease that most of our knowledge of triplet repeat expansion is derived.

Triplet Repeat Expansion Disorders

One manifestation of DNA slippage involving simple repetitive sequences is the instability of certain trinucleotide repeat sequences. This mutational mechanism underlies fragile X mental retardation syndrome, a condition associated with the presence of a fragile site on the X chromosome (FRAXA). The brain-expressed FMR1 (Xq27.3) gene responsible contains an unusual (CGG)n repeat in its 5' untranslated region. This repeat exhibits copy number variation of between 6 and 54 in normal healthy controls, between 52 and >200 in phenotypically normal transmitting males (the 'premutation') and between 300 and >1000 in affected males (the 'full mutation'). Thus a continuum exists between a copy number polymorphism present in the general population, the asymptomatic premutation which involves limited expansion of (CGG)n copy number, and the full mutation which appears to require copy number expansion beyond a certain threshold value. Expansion of premutations to full mutations is thought to be a prezygotic event and occurs only during female meiotic transmission. Alleles with a repeat copy number of <46 do not exhibit elevated meiotic instability. By contrast, for alleles with 52-113 repeat copies, the premutation expands to the full mutation in 70% of transmissions whereas the corresponding figure for alleles with >90 repeat copies is 100%. The probability of repeat expansion thus correlates with the repeat copy number in the premutation allele, consistent with a mechanism of slipped mispairing during replication. Expansion of a sequence can thus itself lead to further expansion, a process termed 'dynamic mutation' by Richareds et al. (1992). In FRAXA, triplet repeat expansion is thought to exert its pathological effects by down-regulation of FMR1 gene expression through hypermethylation of the promoter region upstream of the CGG repeat and repression of translation of the FMR1 transcript.

The discovery of this novel mutational mechanism has led to the recognition that the expansion of unstable triplet repeats is also responsible for a number of other human inherited diseases. These often manifest a wide range of clinical severity and possess unusual features such as increasing severity and penetrance in successive generations ('anticipation') and a sex bias in the transmission of the disease which correlates with the degree of meiotic instability and allelic expansion. The nature and location of the repeat sequence involved varies between disease states as does the extent of the expansion necessary to bring about symptoms of disease, and of course the mechanism of pathogenesis consequent to repeat expansion. It should however be noted that repeat expansion is not confined to triplet repeats; minisatellite expansion can also occur as in progressive myoclonus epilepsy and the clinically asymptomatic fragile sites FRA16A and FRA16B.

Triplet repeat expansions have so far been found to be associated mainly with three types of sequence: CAG (with complement CTG), CGG (with complement CCG) and GAA (with complement TTC). Sequence specificity implies a role for DNA secondary structure in the expansion mechanism. Such repeats are known to form stable hairpin loop structures which become more stable with increasing repeat number. DNA polymerase progression appears to be blocked by CTG and CGG repeats and the resultant idling of the polymerase may serve to catalyze slippage leading to repeat expansion.

A number of diseases are characterized by CAG repeat expansion within the gene coding region. This is thought to constitute a gain-of-function mutation through the incorporation of a polyglutamine tract into the protein product. Several factors are known to influence the stability of triplet repeats viz. the type of sequence, length of repeat, whether the repeat is interrupted or not, and the orientation of the repeat relative to the origin of replication. The removal of interrupting point mutations from repeat arrays is thought to be very important in promoting triplet repeat expansion and is significant in an evolutionary context as well as in cases of disease. Removal of these point mutations could occur simply by single base-pair substitution or by gene conversion or unequal crossing over. Other inherited conditions result from expansion of a triplet repeat in the 3' untranslated region (myotonic dystrophy; DMPK) and an intron and the resulting mechanisms of disease are consequently different.

Nature and Distribution of Triplet Repeats in the Human Genome

Trinucleotide repeats are 10 to 100-fold less frequent than (AC)n repeats and different types of trinucleotide repeat occur at different frequencies. Thus repeats (AAT)n and (AAC)n are the most frequent trinucleotide repeats found in the human genome, with (AAT)n exhibiting a high degree of copy number polymorphism. Both (CAG)n and (CCG)n repeats appear to be over-represented in the human genome, the former being polymorphic in number in the genomes of humans and non-human primates. (CAG)n repeats are rare in intronic regions, possibly because of their similarity to the acceptor splice site consensus, CAGG. Long homopeptides are present in 1.7% of human protein-coding sequences.

A number of human genes contain polymorphic triplet repeats (e.g. cadherin 2, (CDH; 18q12), breakpoint cluster region (BCR; 22q11), glutathione-S-transferase (GSTA1; 6p12), Na^{+}/K^{+} ATPase β1-subunit (ATP1B1; 1q22-q25) but these repeats are not known to exert any pathological effect. Other genes have been identified solely on account of their possession of polymorphic trinucleotide repeats and these genes represent candidate loci for involvement in complex diseases. Trinucleotide repeat containing genes are also found in other species, including mouse, but no nonhuman example of triplet repeat expansion as a cause of a genetic disease has yet been documented.

Replication slippage involving short GC-rich motifs ('expansion segments') has occurred during the evolution of the vertebrate genes encoding the 28S and 18S ribosomal RNAs. Interestingly, different segments of the 28S rRNA subunit gene appear to have coevolved by 'compensatory slippage' allowing RNA secondary structure to be conserved as a consequence of runs of sequence motifs in one region being compensated for by complementary motifs in another.

Origin of Expanded Triplet Repeats

In myotonic dystrophy, all Caucasian and Japanese DM chromosomes possess a specific haplotype, whilst a second disease-associated haplotype has been found in Africans. Similarly, haplotype analysis has pointed to a single origin for Japanese and Caucasian

Machado-Joseph disease chromosomes, a single origin for Japanese and Caucasian dentatorubral and pallidoluysian atrophy chromosomes, a single origin for the Friedreich ataxia expansion, at least two origins for the Huntington disease expanded repeat and a small number of FRAXA progenitor chromosomes.

One of the best examples of the evolutionary emergence of an expanded triplet repeat is that found in the SRP14 gene (15q22) encoding the 14 kDa *Alu* RNA-binding protein. The human protein is larger than that of mouse and dog on account of an extra 28 residue alanine-rich C-terminal tail which is translated from a 3' GCA-rich trinucleotide repeat. In the prosimian *Galago*, the relevant sequence at the site of the human triplet repeat is GCA GCA, whereas the mouse possesses the sequence CCA GCA. By contrast, the African green monkey possesses 33 GCA repeats, whilst the owl monkey possesses 52 suggesting that a CCA→GCA substitution occurred in an ancestral prosimian thereby creating two consecutive GCA codons which then facilitated GCA expansion in higher primates. Interestingly, however, no intra-specific variability in repeat size was detected in primates.

Evolution of Repeat Number in the Genes Underlying Disorders of Triplet Repeat Expansion

The normal range for CAG repeat number in the HD gene is 8-35. Although modal CAG repeat length is fairly similar between different human populations, the degree of spread varies, being greatest among Africans and lowest among the Japanese. The breadth of this normal range suggests that natural selection is acting weakly if at all on HD alleles below the disease threshold. However, the population distribution of CAG repeat number in the HD gene exhibits an apparent asymmetry in that more alleles lie above rather than below the modal length. Using computer simulations, Rubinsztein et al. (1994) have shown that the distribution of HD alleles in human populations is explicable in terms of a simple length-dependent mutational bias. The observed distribution of alleles is thus explicable merely in terms of mutation and genetic drift thereby obviating the need to invoke positive selection. It may be however, that coding sequences can tolerate CAG repeat-encoded polyglutamine tracts relatively well, thereby minimizing the effect of negative selection. More controversially, in the case of myotonic dystrophy and Machado-Joseph disease, *meiotic drive* (also termed *segregation distortion*), the excess recovery of one of a pair of alleles in the gametes of an heterozygous parent, has been proposed as being responsible for maintaining the frequency in the population of chromosomes bearing triplet repeats capable of expansion into the disease range.

Djjan et al. (1996) examined the disease-associated CAG repeats in the HD, MJD, AR, and SCA1 orthologues of various nonhuman primates species. For the HD and MJD genes, CAG copy number was polymorphic in the nonhuman primates but there was essentially no overlap with the normal human range, findings also reported by Limprasert et al. (1996). The AR gene was also found to be polymorphic in nonhuman primates but with minimal overlap with the normal human range. For the SCA1 gene, the average repeat copy numbers exhibited by *Pan*, *Gorilla*, *Hylobates*, *Macaca*, and *Cercopithecus* were within the normal human range but were lower than the modal number is lower in nonhuman primates than in humans. From what we know of the propensity of CAG repeats to expand, this is much more likely to be due to a higher rate of expansion in humans than the alternative: contraction in the other primate species. In the case of the HD gene, this interpretation is supported by the finding of a similar number of CAG repeats in the murine HD gene to that found in the nonhuman primate HD genes. The high number of CAG repeats in the human HD gene is therefore due primarily to expansion within the human lineage. This contrasts with the expansion of CAG number in the AR gene which began earlier in hominoid evolution; the great apes contain an expanded CAG repeat not found in rodents. Thus, the AR CAG repeat expansion began prior to ape-human divergence but continued in human after divergence from the great apes. The most dramatic example of CAG expansion is however that of the involucrin (IVL; 1q21) gene which is so rich in CAG repeats and codons derived from CAG that it is likely that it is descended from a simple poly (CAG) sequence.

The CGG repeat in the 5' UTR of the FMR1 gene has been studied in 44 mammalian species form 8 different orders. The presence of this repeat in all species examined indicates that the CGG repeat has been conserved for over 150 Myrs and is therefore likely to be have some functional significance. Repeat length was found to be similar among the 24 nonprimate species, ranging from 4 to 12 units (mean 8.01 ± 0.8). By the contrast, the mean length of the repeat among the 20 primate species examined was 20.1 ± 2.3. Copy number polymorphism was not found to be limited to human, with *Ornythorhyncus* (platypus), *Artibeus* (phyllostomid bat) and *Pan* (chimpanzee) all possessing polymorphic repeats. Parsimony analysis predicted that the early mammalian CGG repeat was short (4-9 units) and uninterrupted, whereas an increase in copy number beyond ~20 repeats appears to have occurred at least three times independently in the Catarrhini. These expansions in the hylobatid apes, great apes and the cercopithecoid monkeys were associated with the addition of specific interspersions, CGA, AGG and CGGG respectively. These interspersions are unlikely to have arisen through DNA polymerases slippage and may have been mediated by unequal crossing over or gene conversion.

A comparison of trinucleotide repeats in human versus rodent coding sequences has indicated that, by and large, orthologous repeats have not been conserved for long periods of evolutionary time, either in terms of their size or location. As yet, there are no known pathological equivalents of human dynamic mutations in other species but this may merely reflect bias of ascertainment. Thus, nonhuman examples of disease-associated triplet repeat expansions may not yet have come to our attention and it may be that these expansions have no counterpart in the human orthologues. Similarly, the observation that triplet repeats associated with human disease are sometimes found to be both polymorphic and expanded beyond the mammalian ancestral state in non-human primates may also be viewed in terms of a bias of ascertainment rather than any intrinsic propensity of triplet repeats to expand in primate genomes. Finally, noting that most of the expanded triplet repeats are found in genes expressed in the nervous system, Hancock (1996) proposed that triplet repeat expansion may have occurred in parallel with the increase in brain complexity so characteristic of the human lineage and may even have been instrumental in bringing about this increase in complexity.

Unique Case of Involucrin

Involucrin is a unique protein in that it has evolved very rapidly in the primate lineage through changes in the length and copy number of tandem repeats within the coding region with the result that a full two thirds of its coding region has been created within the anthropoid lineage. The evolution of this gene/protein is therefore worth examining in some detail.

Involucrin is the most abundant protein component of the keratinocyte envelope and in cross-linked to membrane proteins by a transglutaminase. The coding region of the human involucrin (IVL; 1q21) gene lies within a single exon and encodes a 585 amino acid glutamine-rich protein. The human IVL gene is extremely rich in CAG repeats (encoding glutamine) and codons derived from it (CAG and CTG). Thus, CAG codons, and codons one substitution removed from

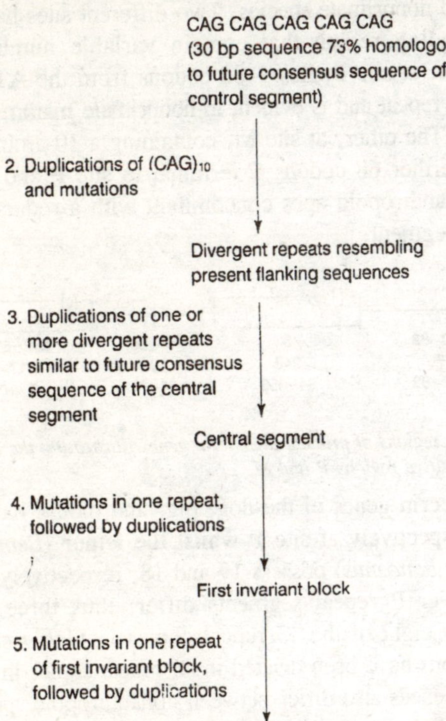

Fig. 5.11. Proposed scheme for the evolution of the human involucrin (IVL) gene.

CAG, comprise 64% of the total codon number. It is possible therefore that this gene is descended from a simple poly(CAG) sequence which has been subsequently modified by nucleotide substitution.

The time of origin of the involucrin gene is unclear but the gene is present in a wide variety of mammals and may be present in all terrestrial vertebrates. The length of the involucrin molecule has continued to grow by successive addition of repeats from prosimians (average 409 residues) through New World monkeys (average 488 residues) and Old World monkeys (average 514 residues) to hominoids (average 632 residues). The coding region outside the segment of repeats has by contrast changed very little in length. Since glutamine residues serve as amine acceptors in the transglutaminase-catalyzed cross-lining, the primordial glutamine-rich involucrin may have been able to function as a substrate for transglutaminase.

Involucrin gene sequences have now been determined in a range of primate and nonprimate species. Two different sites have been found within the coding region that contain variable numbers of repeat segments. One, at site P, lies ~80 codons from the ATG, contains a 16-amino acid repeat and is evident in nonprimate mammals, prosimians and tarsioids. The other, at site M, containing a 10-amino acid repeat, is located a further 68 codons C-terminal to site P and has increased in size in the anthropoid apes concomitant with a reduction in size of the P repeat segment.

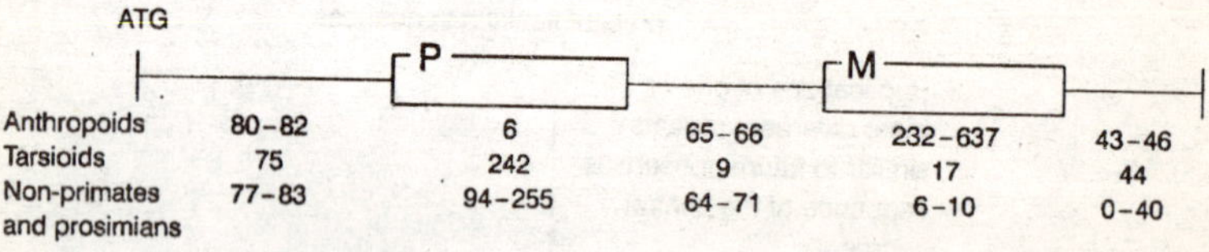

Fig. 5.12. Coding regions of primate involucrin genes illustrating the relative locations of repetitive regions P and M.

The involucrin genes of the dog, pig, and mouse have 6, 13, and 21 repeats, respectively at site P whilst the lemur (*Lemur catta*) and tarsier (*Tarsius bancanus*) possess 19 and 18, respectively. Within each species, however P repeat segments differ; thus three codons have been deleted in eight of the 13 repeat copies in *Galgo crassicaudatus* whilst two codons have been deleted in all repeat copies in *T. bancanus*. Consensus sequences also differ between nonanthropoid species in terms of nucleotide differences that occurred in any position in a codon. Changes in a consensus nucleotide must have arisen from the occurrence of a nucleotide substitution in one repeat followed by the correction of the analogous nucleotides at the corresponding positions in the other repeats, with adjacent repeats tending to be more homogeneous. Green and Djjan (1992) have proposed that a form of intragenic gene conversion may have been responsible for this phenomenon. Although the P segment was retained by prosimians and tarsioids, it was modified by site specific deletions, the addition of certain repeats and nucleotide substitutions which served to alter the consensus codons of the repeats by a process of correction operating between neighbouring repeats. It is possible that the retention of the P segment was essential for the function of the protein since repeats at site M were lacking.

In the anthropoid apes, the M segment increased in size concomitant with the dramatic reduction in size of the P repeat segment. Perhaps deletion of the P segment was only possible after a sufficient number of repeats had been generated within the M segment.

Repeat Number	Repeat Type	1	2	3	4	5	6	7	8	9	10	Number of altered codons per repeat
10	A	K	H	L	E	Q	Q	E	G	Q	L	5
9	B	E	L	P	E	Q	Q	V	G	Q	P	10
8	A	K	H	L	E	Q	E	E	K	Q	L	10
7	B	E	L	P	E	Q	Q	E	G	Q	L	3
6	A	K	H	L	E	K	Q	E	A	Q	L	2
5	B	E	L	P	E	Q	Q	V	G	Q	P	3
4	A	K	H	L	E	Q	Q	E	K	Q	L	2
3	B	E	H	P	E	Q	Q	E	G	Q	L	7
2	A	K	H	L	E	Q	Q	E	G	Q	L	6
1	A	K	D	L	E	Q	Q	E	G	Q	L	10
												Mean 5.8

Fig. 5.13. Consensus amino acid sequences for each of the 10 repeats of the early region of the M segment of involucrin.

This notwithstanding the M segment continued to grow by successive addition of repeats from prosimians through New World Monkeys and Old World monkeys to hominoids. However, the repeats have the same consensus sequence in all the anthropoid apes. The M segment has been divided into early, middle and late regions which have been added vectorially in a 3'→5' direction and the repeats from these regions are shared to differing extends by different anthropoid species. Thus, the early region is common to all anthropoids, the middle region is shared by different species but to a lesser extent, and the late region is species-specific and must have developed after the divergence of the different species. After its divergence from the cercopithecoids, the hominoid lineage added 10 repeats to the middle region. After divergence from the common lineage, *Hylobates*, *Pongo*, and *Homo* acquired species-specific repeats in the late region. Another shared repeat segment is the early/middle (e/m) extension 5' to the late region which was probably formed independently in Old World and New World monkeys.

The site of repeat addition, which in the common anthropoid lineage was located in the P segment, moved in a 5' direction during the evolution of both the New World and Old World monkeys after their divergence. Deletions of repeats also took place within the early, middle and e/m regions of the M segment although these were relatively few in number as compared to the additions. These deletions have occurred independently of the vectorial process of repeat addition both spatially and temporally.

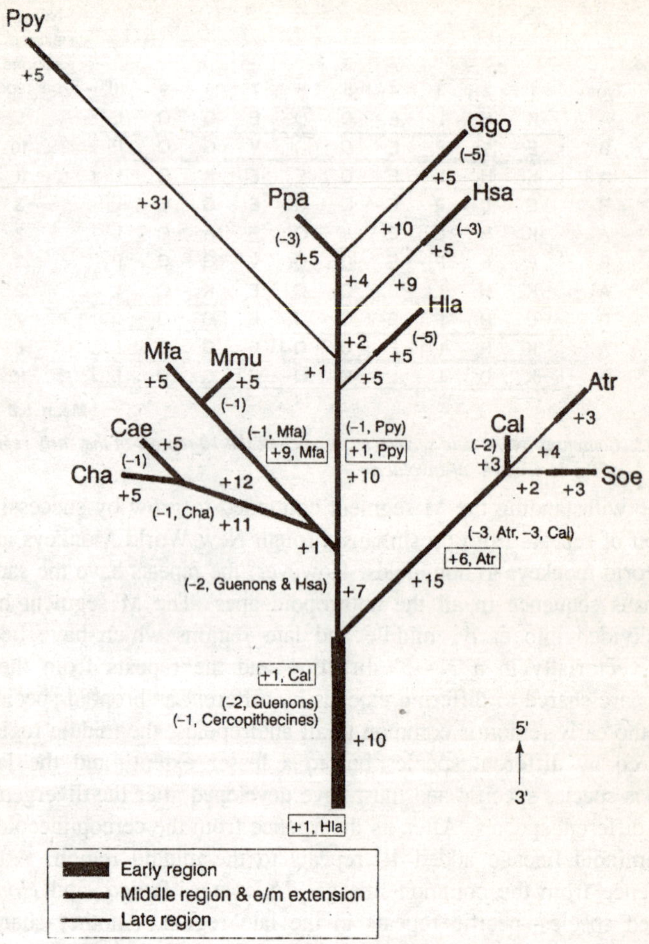

Fig. 5.14. Evolutionary tree of the M region of the primate involucrin gene detailing repeat additions.

In addition to the inter-specific differences in the size of the involucrin molecule, the higher primates also exhibit size polymorphisms which result from variation in the number of repeats in the late region. In humans, the polymorphism comprise variable numbers of B and B^S repeats within the late region. The most common allele in white Caucasians contains nine repeats ($3B^S$, 5B, 1A) but a second 'Mormon' allele ($2B^S$, 5B, 1A) has also been described. Blacks possess the 'Mormon' allele plus three other alleles containing 6, 7, or 9 B repeats. Repeat number polymorphisms in the late region have also been reported in *Aotus trivirgatus*, *Macacca mulatta* and *Gorilla gorilla*. The repeat pattern observed in anthropoid apes is explicable in terms of the presence in the involucrin gene of a hotspot at which repeats are generated by unequal recombination. In the human lineage, the location of this hotspot varies between racial groups such that in whites it is located within the B^S repeat region whilst in blacks, it is within the B repeat region. Thus, although the process of vectorial repeat addition has operated in both blacks and whites, there has been a difference between these racial groups in the sites of repeat addition, a finding that is consistent with an endogenously controlled mechanism.

Attempts have been made to relate the evolution of a larger involucrin molecule with a different repeat structure to the trend in anthropoid apes towards relative hairlessness. The P segment of repeats could conceivably have conferred some selective advantage during the early stages of mammalian and primate evolution as might the M segment when it emerged. At a biochemical level, natural selection might have favoured the retention of repeat-bearing segments in involucrin so as to provide a substrate for transglutaminase. Once the M segment has been generated, the repeats at site P could then have been deleted without selective disadvantage. However, there is a very considerable degree of latitude between different species in terms of what constitutes an effective transglutaminase substrate. In a comparison of the involucrins of 19 mammalian species, only 3.1% of amino acid residues were uniformly conserved. Thus, on balance, neither the very high rate of evolutionary change nor the observed patterns of mutational change in the involucrin gene provide convincing evidence for natural selection being the major motive force behind involucrin evolution in higher primates. Indeed, it is unlikely that natural selection could account on its own for either the highly variable pattern of repeat addition or the numbers of repeats added.

Natural selection and neutral mutation/genetic drift are both processes that govern how mutations, once they have arisen, spread through a population with the possibility of eventually becoming fixed. By contrast, the evolution of involucrin appears to be primarily mechanism-based, being dominated by endogenously controlled and spatially targeted mechanisms of mutagenesis. Such mechanisms are likely to be the major factor responsible for the *directed evolution* of involucrin, directed in the sense that the mechanisms promote change in a specific direction (i.e. increases in repeat number or gene conversion between homologous repeats) without the necessary assistance of natural selection.

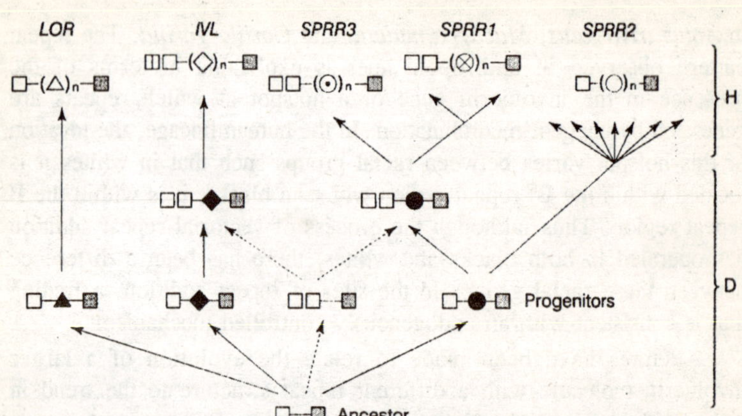

Fig. 5.15. Hypothetical scheme for the evolution of the involucrin (IVL), loricrin (LOR) and small proline-rich protein (SPRR) genes from a single ancestral gene.

Interestingly, several other human genes encoding epithelial proteins are located on the long arm of chromosome 1 [filaggrin (FLG), loricrin (LOR) epithelial mucin (MUC1), and the small proline-rich proteins 1A (SPRR1A), 1B (SPRR1B) 2A (SPRR2A) and 3 (SPRR3)]. These genes share a similar structure, contain multiple tandem repeats and are believed to be evolutionarily related both to involucrin and to each other. The LOR, FLG and MUC1 genes also exhibit repeat copy number polymorphism. The apparent association between chromosomal location and copy number polymorphism almost certainly reflects the possession by evolutionarily related proteins of a repeat structure that is particularly prone to the process of vectorial addition of repeats.

6

MULTIGENE FAMILIES

The development of DNA reannealing and hybridization techniques in the mid-1960s provided a convenient, though crude, means for measuring the degree of divergence between related sequences and for exploring the structure and organization of eukaryotic genomes. Application of these techniques led to two intriguing observations. First, the genome of higher organisms is composed of highly and moderately repeated sequences as well as single-copy sequences. Second, denatured repetitious DNA strands from one species form duplexes more rapidly with one another than they do with homologous strands from a different species. In other words, within one species the members of a repeated sequence family may be highly similar, whereas members from closely related species may differ greatly. Edelman and Gally (1970) found this observation perplexing and suggested a new concept, namely coevolution of DNA sequences, that is, the members of a sequence family in a species evolve together, so that they maintain high similarities among themselves, as they diverge from members in other species.

An excellent example of this phenomenon was provided in a study of the ribosomal DNA from *Xenopus*, the African toads. In *Xenopus* and most other vertebrates, the genes specifying the 18S and 28S ribosomal RNAs are present in hundreds of copies and are arranged in one or a few tandem arrays. Each repeat unit consists of a transcribed and a nontranscribed spacer (NTS) region, which is now commonly called the intergenic spacer (IGS). The IGS regions within the ribosomal DNA of *Xenopus laevis* are highly homogenous but are very different from those in *X. borealis*. The large interspecific

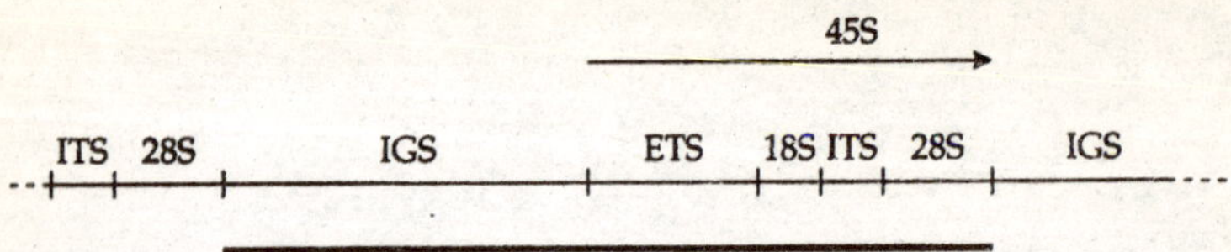

Fig. 6.1. Diagramatic representation of a typical repeated unit of rRNA genes in vertebrates.

difference indicates that the IGS regions accumulate mutations rapidly in evolutionary time. Therefore the high intraspecific homogeneity is maintained not by selective constraints (purifying selection) but by certain mechanisms by which mutations in a repeat can spread "horizontally" to all members in the same family. For this reason, Brown et al. (1972) called this phenomenon *horizontal evolution*, in contrast to vertical evolution, which refers to the spread of a mutation to individuals in a breeding population through generations. Later the terms *coincidental evolution* and *concerted evolution* were suggested. The latter is now commonly used in the literature.

This new phenomenon was so puzzling that many mechanisms have been proposed for its occurrence. These include the saltatory replication hypothesis, unequal crossing-over, replication slippage, gene conversion, and duplicative transposition. The phenomenon has been a subject of intense research in the last two decades. A large body of data obtained by restriction enzyme analysis and DNA sequencing techniques has attested to the generality of concerted evolution in multigene families. Moreover, theoretical studies have greatly sharpened our understanding of the population dynamics of concerted evolution. However, our knowledge is still limited in many aspects, such as in the rate of unequal crossing-over, the rate of gene conversion in higher organisms, and the rate of concerned evolution in a family.

Gene Conversion and Unequal Crossing-Over

Gene conversion and unequal crossing-over are now thought to be the two most important mechanisms for the occurrence of concerted evolution. They are also the two mechanisms that have received extensive quantitative treatments. To better understand their roles in the evolution of multigene families, we discuss some characteristics of gene conversion and unequal crossing-over, especially the rates of occurrence, as they determine the intensity of concerted evolution.

Gene Conversion

Gene conversion is a recombination process in which two sequences interact in such a way that one is converted by the other. It is a nonreciprocal process, because one sequence is changed whereas the other is not.

Fig. 6.2 shows a classification of conversional interactions according to the pathways involved; the chromatids are depicted after DNA synthesis but before segregation. (Chromatids are the copies of a chromosome produced by replication.) Pathway c is the interaction between two alleles at the same locus and is known as classical (allelic) gene conversion because this was the first type of gene conversion studied. The other pathways are (a) intrachromatid, (b) sister-chromatid, (d) semiclassical (nonallelic between homologous chromosomes,), and (e) ectopic (heterochromosomal) conversion events, which refer, respectively, to interactions between genes (a) on the same chromatid, (b) at nonhomologous loci on sister chromatids, (d) at nonhomologous loci on homologous chromosomes, and (e) on nonhomologous chromosomes. Here we assume that gene conversion occurs immediately after chromosome replication, so interactions between genes at homologous loci on sister chromatids have no effect and have not been included in the above classification. Collectively,

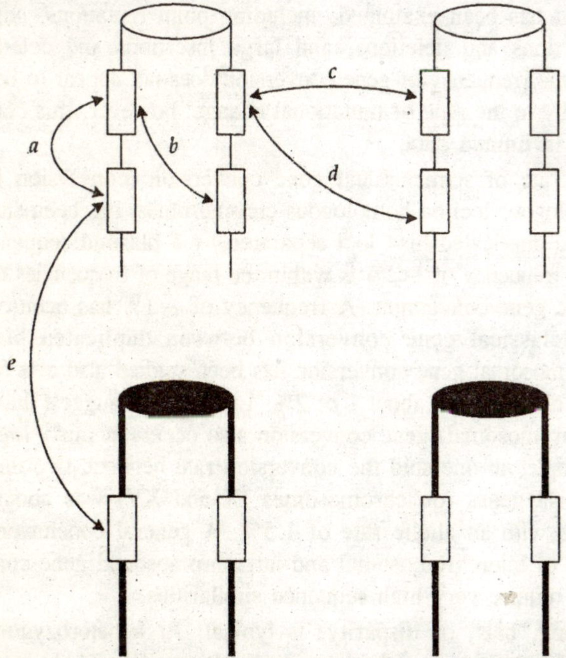

Fig. 6.2. Types of gene conversion between repeated genes.

intrachromatid and sister-chromatid interactions are interchromosomal, whereas classical, semiclassical, and ectopic interactions are interchromosomal.

Gene conversion has been demonstrated in many species of fungi, in *Drosophila*, and in *Zea mays*. However, even in the best studied organism, the yeast *Saccharomyces cerevisiae*, data on the rate and other features of gene conversion are limited. The following brief summary is taken largely from the yeast data reviewed in Petes and Hill (1988) and Petes et al. (1991). It should be cautioned that conclusions drawn from yeast data may not be applicable to other organisms; for example, compared with higher eukaryotes, yeast seems to manifest a high rate of gene conversion, though the observed high rate might be partly due to the ease of detecting gene conversion events in yeast. Four primary conclusions have been drawn from yeast studies.

First, allelic gene conversion has been observed for every locus that has been examined in detail and the frequencies vary from about 0.5% to 18%; an exceptionally high rate (30%) has been observed at the *his4* locus. Conversion events were observed for every type of allele that has been examined, including point mutations, small (1-4 bp) insertions and deletions, and large insertions and deletions. In general, the frequency of gene conversion does not appear to be related specifically to the type of mutational change; however, this conclusion is based in limited data.

The rate of semiclassical gene conversion (conversion between nonhomologous loci on homologous chromosomes) has been studied by using two duplicated his4 loci separated by a plasmid sequence. The observed frequency of ~2% is within the range of frequencies observed for allelic gene conversion. A frequency of ~1% has been observed for semiclassical gene conversion between duplicated his3 loci. Intrachromosomal gene conversion has been studied also at a few loci; the rates observed are about 1 or 2%. Limited data suggest that ectopic (heterochromosomal) gene conversion also occurs at fairly high rates. For example, in one stud the conversion rate between two duplicated 1.7-kb his3 genes (on chromosomes IX and XV) was about 0.5%, compared with an allelic rate of 1.5%. A general conclusion is that high rate of interchromosomal and intrachromosomal gene conversion may not require very high sequence similarities.

Second, bias, or disparity, is typical. At a heterozygous locus with alleles A and a, if the two conversion events 3A:1a and 1A:3a occur with equal frequency, the conversion is said to show parity or no bias. If deviation from parity occurs, the allele whose frequency increases as a result of the deviation is said to have a *conversional advantage*; the other allele has a *conversional disadvantage*. In an examination of the data summarized in Fogel et al. (1978), Nagylaki and Petes (1982) found that gene conversion bias is a common phenomenon and that the degree of bias appears to depend on the type of mutation (i.e., substitution mutation and deletion or insertion). In fact, approximately half of the mutant sites examined in yeast showed sufficiently large deviations from parity to be detected with a sample size ≥ 100. For small insertions (4 bp), the studies of Borts and Haber (1987) and Symington and Petes (1988) showed parity or a small bias. By contrast, the studies of Fogel et al. (1981), McKnight et al. (19?1), and Pukkila et al. (1986) indicated that large (≥ 100 bp) deletions frequently show departures from parity.

Third, the amount of DNA transferred during a single gene-conversion event varies from a few base pairs to greater than 12,000 bp. Studying closely linked sites within the ARG4 gene, Fogel and Mortimer (1969) concluded that the tract length of a typical conversion event is several hundred base pairs. In studies involving strains with multiple heterozygous restriction sites, the average tract lengths estimated were: 1500 bp and 3700 bp. Occasionally, much larger conversion tracts were observed in conversion events. The size of the tract depends on the stain and the position in the genome, but within a tract gene conversion is usually continuous.

Fourth, the rate of gene conversion varies from site to site and is context-dependent. Symington and Petes (1988) found that among 82 tetrads associated with a crossover in the 22-kb interval between LEU2 and CEN3, one heterozygous marker converted in 27 tetrads whereas another marker converted only in 4 tetrads. Large (38-fold) differences in the rate of gene conversion were observed between leu2 heteroalleles inserted at different positions in the genome. In *S. pombe* a G→T mutation within the coding region of the ade6 gene stimulates both meiotic gene conversion and crossing-over. The effect of this mutation is context-dependent; when a 3-kb fragment containing the mutant sequence was placed in a different chromosomal location, no stimulation of meiotic recombination was observed, suggesting that some nucleotides more than 1 kb from the mutation were required for hotspot activity.

Unequal Crossing-Over

Unequal crossing-over creates a sequence duplication in one chromosome (or chromatid) and a corresponding deletion in the other;

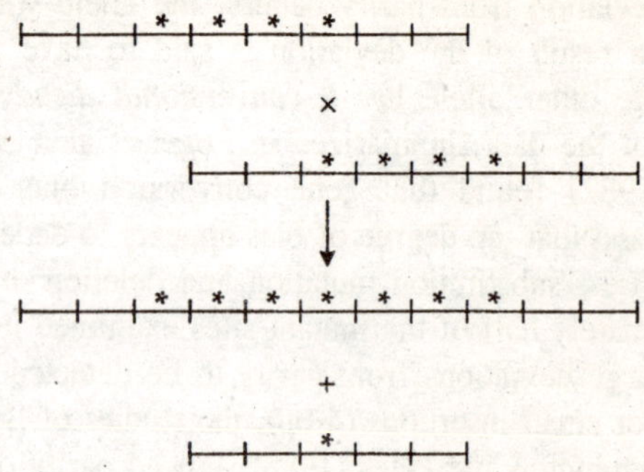

Fig. 6.3. Model of unequal crossing over.

thus, unlike gene conversion, it is a reciprocal recombination process. Fig. 6.3 shows an example in which an unequal crossing-over event has led to duplication of three repeats in one daughter chromosome and deletion of three repeats in the other. Obviously, unequal crossing-over can produce fluctuations in the number of repeats in a family. One good example is the rRNA genes in *Drosophila*. A large deletion of rRNA genes can result in a "bobbed" mutant, which, in homozygous form, manifests small bristles, slow development, reduced growth and fertility, and poor viability. Conversely, a bobbed mutant can revert to normal by an increase in the number of rRNA genes.

Unequal crossing-over may occur between the two sister chromatids of a chromosome or between two homologous chromosomes at meiosis. The latter type apparently can occur with high frequency. For example, a frequency of 17% was observed in a study with two duplicated his4 loci separated by a bacterial plasmid sequence; this frequency is much higher than that (2%) of semiclassical gene conversion for the same locus. The frequency (~1.5%) of unequal sister-chromatid (intrachromosomal) crossing-over is much lower than that of unequal crossing-over between homologous chromosomes but is as high as that of intrachromosomal gene conversion. A similar frequency of unequal sister-chromatid crossing-over was also observed by Malony and Fogel (1987). Unequal crossing-over may also occur between nonhomologous chromosomes; however, except for some cases such as that of rRNA genes in man and apes, this type of crossing-over is not important because it creates translocations, which generally have deleterious consequences.

Examples of Concerted Evolution

Numerous examples of concerted evolution of multigene families have been proposed, including the 5S DNA family in *Xenopus*, the γ-globin genes in primates, the α-globin genes in primates, the heatshock genes in *Drosophila* and the chorion gene superfamily in the silkmoth. In the following we consider three examples that have been studied in detail.

X-linked Colour Vision Genes in Higher Primates

As noted previously, Old World monkeys (OWM), apes, and humans are trichromatic because they possess three colour photopigments: the short-wave (blue), middle-wave (green), and long-wave (red) pigments. The gene encoding the blue pigment is autosomal, whereas the genes for the red and green pigments are tandemly arrayed on the X chromosome. The red and green pigment genes are believed to have arisen from a duplication event before the separation of the human-ape lineage from the OWM lineage, though after the separation of the OWM-ape-human lineage from the New World monkeys, which posses only one X-linked and one autosomal pigment gene. Indeed, the tree constructed from the synonymous substitutions in exons 4 and 5 of the two genes from a human and a baboon (an OWM) is consistent with the view that the two genes diverged before the separation between the human and OWM lineages. The degrees of synonymous divergence in exons 4 and 5 are 8.1% between the two human genes and 11.5% between the two baboon genes. These values are slightly higher than the average divergence between human and Old World intron sequences (7.1%) and so are consistent with the above proposed duplication date.

Given the strong evidence that the red and green pigment genes diverged before the human-OWM split, one would expect the homologous

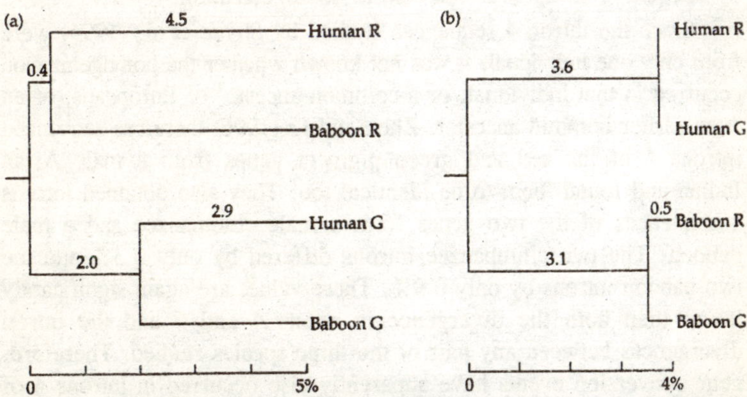

Fig. 6.4. (a) Tree inferred from exons 4 and 5 of the red (R) and green (G) pigment genes of human and baboon. (b) Tree inferred from intron 4 sequences of human and baboon red and green pigment genes.

introns from the two genes to have diverged to a degree similar to the average divergence between human and OWM introns (7.1%). However, the intron 4 sequences (each ~2-kb long) from the red and green genes of a normal human were found to be identical. This absence of divergence was totally unexpected, because the two genes should have diverged more than 20 million years ago, a minimum estimate for the human-OWM split. At any rate, it is significantly lower than the nonsynonymous divergence (2.0%±0.5%) between the two genes from the same human individual studied, contrary to the usual situation in which introns are less conserved than exon. A simple explanation for this observation is that the intron 4 sequences between the two genes have been completely homogenized by gene conversion. This observation is difficult to explain by unequal crossing-over, because neither exon 4 nor exon 5 has been homogenized. In addition, the intron 2 sequences from the two genes were found to differ by only 0.3%, indicating that this intron has been almost completely homogenized by gene conversion.

In the above discussion, we have simplified the presentation by neglecting the fact that a normal human X chromosome may carry more than one red or one green pigment gene. The multiple copies of the green (or red) pigment genes in humans are, however, highly similar, which can be a result of either recent origins or homogenization or both. In this case, homogenization can occur by either gene conversion or unequal crossing-over—obviously, the variation in gene copy number should have resulted from unequal crossing-over. As will be explain later, contraction and expansion in copy number in a sequence family by unequal crossing-over can lead to homogenization.

Since the intron 4 sequences studied by Shyue et al. (1994) were from only one individual, it was not known whether the homogenization occurred in that individual, or a common ancestor of Europeans, or an even earlier hominid ancestor. Zhou and Li (1996) therefore sequenced introns 4 of the red and green pigment genes from a male Asian Indian and found them to be identical too. They also obtained introns 4 sequences of the two genes from a male chimpanzee and a male baboon. The two chimpanzee introns differed by only 0.3% and the two baboon introns by only 0.9%. These values are again significantly lower than both the divergence in exons 4 and 5 and the intron divergences between any pair of the three species studied. Therefore, gene-conversion events have apparently also occurred in introns 4 of the two genes in chimpanzee and in baboons. For this reason, the tree inferred fro the human and baboon intron 4 sequences groups the two human introns in one cluster and the two baboon introns in another, and is drastically different from that inferred from exons 4 and 5.

It is possible that some of the putative gene-conversion events in introns 4 also involved either or both of exons 4 and 5, which are only about one-tenth the length of intron 4. In addition, there is evidence for gene-conversion events in exons in humans and OWMs. Nevertheless, exons 4 and 5 remain relatively divergent between the two genes. A most likely explanation of these observations is that natural selection for maintaining the distinct functions of exons 4 and 5 of the red and green pigment genes has acted against sequence homogenization in either of these exons. Homogenization of exon 4 or 5 between the red and green pigment genes would reduce the spectral sensitivity differences between the two genes, and this would be disadvantageous to its carriers, so that the mutant haplotype would eventually be eliminated from the population.

In summary, a lesson from the X-linked colour pigment gens in higher primates is that concerted evolution may occur with different intensities in different regions of the genes, owing to different selective constraints in different regions.

Ribosomal RNA Genes in Humans and Apes

The rDNA multigene family in humans and apes provides a model system for studying the concerted evolution of a family of repeated genes distributed on several chromosomes. In humans and chimpanzees, the approximately 400 rDNA genes are distributed among nucleolus organizers located on five pairs of chromosome. As noted, each rDNA repeat unit contains a nonconserved region, i.e., the intergenic spacer (IGS). If rDNA genes on nonhomologous chromosomes have experienced no genetic interaction, each nucleolus-organizer region would have evolved independently, and there should be high variation in the IGS region among different nucleolus organizers within individuals of a species. Thus, a high degree of homogeneity in this region among genes in different nucleolus organizers would suggest that the rDNA family has evolved in a concerted fashion.

Using restriction enzyme analysis, Arnheim et al. (1980) found that different regions of the rDNA repeat unit in humans and the apes have different characteristic evolutionary patterns. The transcribed segments have undergone few changes because the 18S and 28S RNA gene sequences have been highly conserved in evolution. In contrast, differences exist among species in the IGS region with no comparable variation existing within or among individuals of a species.

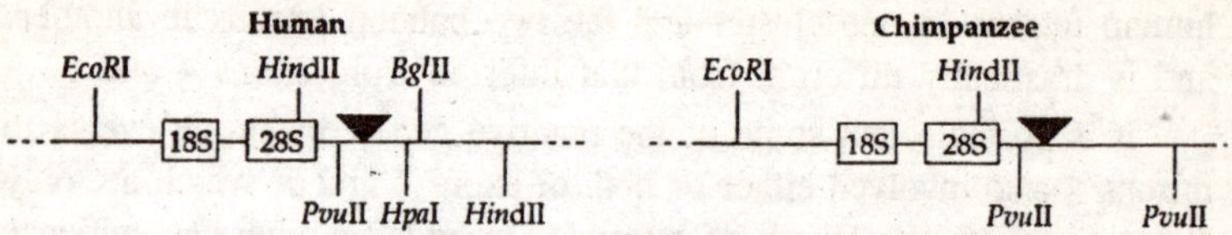

Fig. 6.5. Restriction sites in human and chimpanzees ribosomal RNA genes.

Fig. 6.5 shows a summary of the data in Arnheim et al.'s (1980) analysis of human and chimpanzee rDNA genes. Each human repeat unit has a HpaI site in the IGS region 3' to the 28S gene. The chimpanzee lacks this site. The fact that none of the other great apes (gorilla, orangutan, and gibbon) have a HpaI site in this position suggests that the most recent common ancestor of humans and chimpanzees also lacked this site. Therefore, the HpaI site most probably originated in the human lineage after the human-chimpanzee divergence and eventually became fixed in every human rDNA repeat. Some other restriction sites (e.g., HindII) in the IGS region also demonstrate species-species homogeneity. These observations indicate that in spite of the fact that the rDNA genes are distributed on five pairs of nonhomologous chromosomes, the whole family has evolved together in a concerted manner.

Krystal et al. (1981) have provided further evidence for genetic exchange between rDNA clusters on nonhomologous chromosomes. They examined the distribution of restriction sites that are polymorphic within humans. Their results reveal a general uniform distribution of all of the polymorphisms over the chromosomes (with a few exceptions), and are consistent with a model of concerted evolution that involved exchanges between sister chromatids or homologous chromosomes and between nucleolus organizers on nonhomologous chromosomes. The exchanges could have been either gene conversions or unequal crossing-overs. The latter type of events would cause the number of rDNA genes per cluster to fluctuate randomly. In fact, estimates of the number of rDNA genes on the human nucleolus containing chromosomes by *in situ* hybridization have revealed extensive copy-number variation among the chromosomes within an individual and among the same chromosomes from different individuals.

In humans (and apes) unequal crossing-over events between rDNA genes on nonhomologous chromosomes would probably cause on significant deleterious effects because the rDNA genes are located on the stalks between the short arm and satellite body of chromosomes 13, 14, 15, 21, and 22. For example, Fig. 6.6 shows that an unequal crossing-over event between rDNA genes on chromosomes 14 and 21

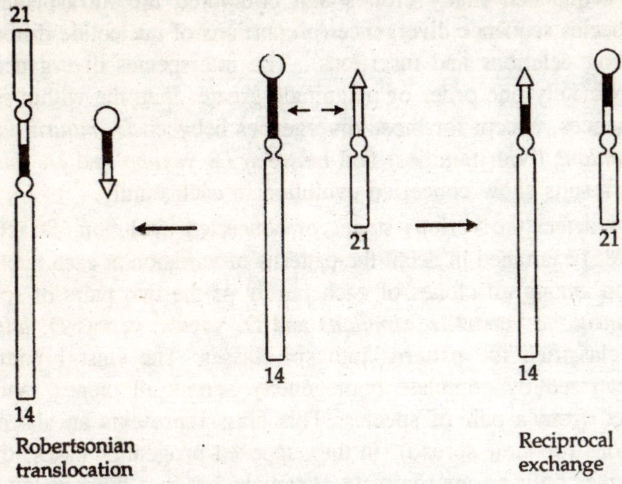

Fig. 6.6. Cytogenetics consequences of genetic exchanges between human rDNA sequences on nonhomologous chromosomes.

produces a translocation involving only rDNA genes and satellite material. Of course, genetic exchange between rDNA genes on nonhomologous chromosomes can also occur via gene conversion. The relative importance of gene conversion and unequal crossing-over in the concerted evolution of rDNA genes in humans and apes cannot be evaluated from the data available at the present time.

The "360" and "500" Nongenic Repeat Families in *Drosophila*

The "360" and "500" repeat families are two abundant tandem families of non-coding sequences found within a group of eight sibling species of the *melanogaster* species subgroup of *Drosophila*; their names signify the approximate lengths (~360 and ~500 bp) of their repeat units. Strachan et al. (1985) chose these two families to illustrate the transition stage of concerted evolution for three reasons. First, the repeats of these two families do not seem to be subject to any discernible natural selection, so that the dynamics of spread of mutant variants due to concerted evolution can be separated from that due to natural selection. Second, their phylogenetic distribution allows a detailed examination of the various stages of concerted evolution. Third, the two families have been evolving in parallel and so their patterns of concerted evolution can be contrasted.

Strachan et al. (1985) randomly selected clones of members of each family from each of the following species: *Drosophila melanogaster*, *D. mauritiana*, *D. simulans*, *D. orena*, *D. teissieri*, and *D. yakuba*.

They sequenced many clones and computed the intraspecies and interspecies sequence divergence (proportions of nucleotide differences including deletions and insertions). The interspecies divergences are approximately one order of magnitude greater than the within-species divergences, except for those divergences between *D. mauritiana* and *D. simulans* (both families) and between *D. yakuba* and *D. teissieri*. These results show concerted evolution in each family.

To detect the various stages of concerted evolution, Strachan et al. (1985) examined in detail the patterns of variation at each nucleotide position among all clones of each family of the two pairs of species: *D. mauritiana* versus *D. simulans* and *D. yakubà* versus *D. teissieri*. They classified the patterns into six classes. The class-1 pattern is characterized by complete homogeneity across all clones randomly sampled from a pair of species. This class represents an absence of mutations (or their spread), in the supposed progenitor bases; this is, of course, only an approximate statement because homogeneity in a limited sample of sequences does not imply homogeneity in the population. Class-2 represents rare mutations (or low levels of subsequent spread), resulting in the appearance of a minority of clones with a new mutation in a given position in one species while the other species remains homogeneous for the progenitor base at the corresponding position. Class-3 covers those cases in which no decision can be made between minority and majority frequencies in that a mutation and the progenitor base are in approximately equal frequencies, while the other species is homogeneous for one of the two bases. No examples of this were found in the 360 family, although such examples were found in

(a)

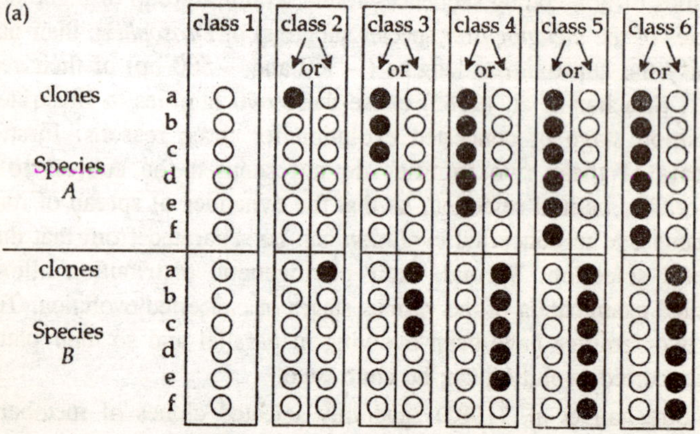

Fig. 6.7. Graphic representation of transition stages during the spread of new mutations.

the 500 family. Class-4 includes those positions in which a mutation, which is apparently absent in one of the species, has repeated the progenitor base in the majority of members in the other species. This interpretation assumes that the base that is in the minority in one species and that is homogeneous in the other species is the progenitor base. Class 5 represents positions in which the to species are internally homogeneous for bases that are diagnostically different for each of the species. This class represents the classical observation of concerted evolution, i.e., the final outcome of the dual process of intrafamily homogenization and population fixation of mutations that arose independently in one or both species. All subsequent mutations beyond this point are represented by class 6.

First, for each of the two families, a greater number of class-5 and -6 nucleotide positions (representing a more extensive spreading of variant repeats) is found between *D. yakuba* and *D. teissieri* than between *D. mauritiana* and *D. simulans*. For example, for the 360 family there are 55 class-5 positions between *D. yakuba* and *D. teissieri* but only 5 class-5 positions between *D. mauritiana* and *D. simulans*. This observation is consistent with the assumption of a relatively constant rate of concerted evolution because *D. yakuba* and *D. teissieri* diverged earlier than did *D. mauritiana* and *D. simulans*.

Second, the rate of spread of mutations is apparently faster in the 360 family than in the 500 family in all four species. For example, for the pair of *D. yakuba* and *D. teissieri*, the proportion of class-5 nucleotide sites is 27.5% for the 360 family but only 6.8% for the 500 family. This difference in rate is expected because the 360 family is confined to the X chromosome, whereas the 500 family members are distributed on all four pairs of chromosomes.

The major mechanism for the concerted evolution in these two families has been suggested to be unequal crossing-over, because this is in keeping with the experimental proof of unequal exchanges within tandem arrays of rDNA that are embedded between some of the arrays of the 360 and 500 families on the X and Y chromosomes. However, there were indications that gene conversions have also contributed to the concerted evolution of these two families.

Concerted Evolution by Gene Conversion

Edelman and Gally (1970) appear to have been the first to suggest that the gene conversion can homogenize repeated genes in a genome. Birky and Skavaril (1976) initiated a theoretical investigation of this problem by using computer simulation. Later, Ohta (1977) used an analytic method (the diffusion theory) to study the extinction time of a variant repeat. The mathematical theory of gene conversion has since been greatly extended; e.g., Ohta (1984, 1985, 1986), Slatkin (1986), Walsh (1986, 1987, 1988), Kaplan and Hudson (1987), Basten and Weir (1990), and Nagylaki (1990). We consider only some simple situations here.

To understand how gene conversion can homogenize repeated genes, let us start with the case of tandem repeats and follow the evolution of a single chromosome lineage. If the family consists of only two repeats, then a single conversion event can result in homogeneity. In a larger family the situation is more complicated, but repeated cycles of gene conversion will eventually lead to homogeneity, if there is no mutation. Nagylaki and Petes (1982) developed a theory for the fixation probability and the mean time to fixation or loss of variant repeat.

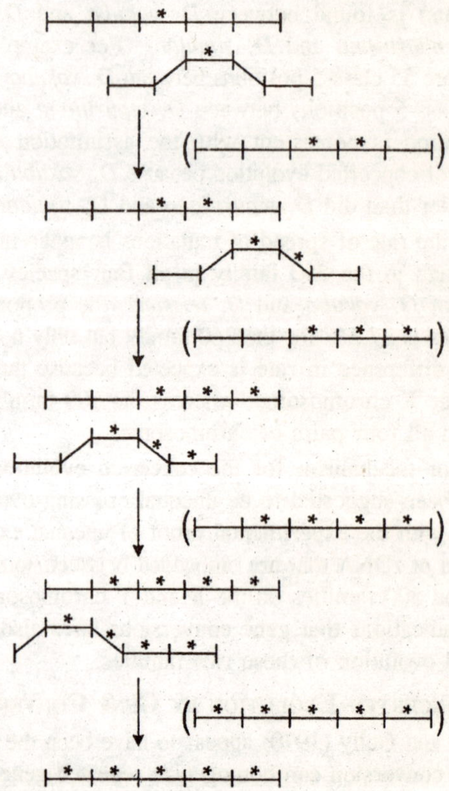

Fig. 6.8. Concerted evolution by gene conversion.

They showed that a small conversional bias (advantage or disadvantage) can have a dramatic effect on the fixation probability of a new variant and that gene conversion can act sufficiently rapidly to be an important mechanism for producing and conserving sequence homogeneity.

The above treatment neglected random genetic drift and the effects of mutation and crossing-over. To incorporate these factors, we assume that the population is diploid and mates randomly, with effective size N, that every mutation creates a new allele (the rate of mutation is u per gene per generation), and that (equal) crossing-over is independent of gene conversion and at most one crossover occurs per generation in the entire multigene family in an individual.

Let us consider the case of interchromosomal gene conversion, which is mathematically simpler than that of intrachromosomal gene conversion. We assume unbiased gene conversion because conversion bias complicates the mathematics greatly. The family consists of n tandem repeats and all loci are equivalent and exchangeable.

In the presence of mutation, complete homogeneity can no longer be attained, but there will be a balance between the homogenizing effect of gene conversion and the diversifying effect of mutation. Following Ohta (1982, 1983a, b), we consider three probabilities of identity: (i) f, the probability that two genes at the same locus, chosen at random from two distinct gametes, and identical; (ii) g_1, the probability that two distinct genes on the same randomly chosen gamete are identical; and (iii) g_2, the probability that two genes at different loci on different gametes, chosen at random, are identical; subscripts 1 and 2 signify whether one or two gametes (chromosomes) are involved. Thus, f represents the expected homozygosity at a locus and h=1-f, the expected heterozygosity, is a measure of intralocus genetic variability in the population; g_1 is an index of homogeneity between repeats within a chromosome; and g_2 incorporates both intralocus and interlocus variation Nagylaki (1984a) showed that a equilibrium the above probabilities of identity are given by

$$\hat{f} = 2(nu + \lambda) / (nD) \qquad \text{...(1a)}$$

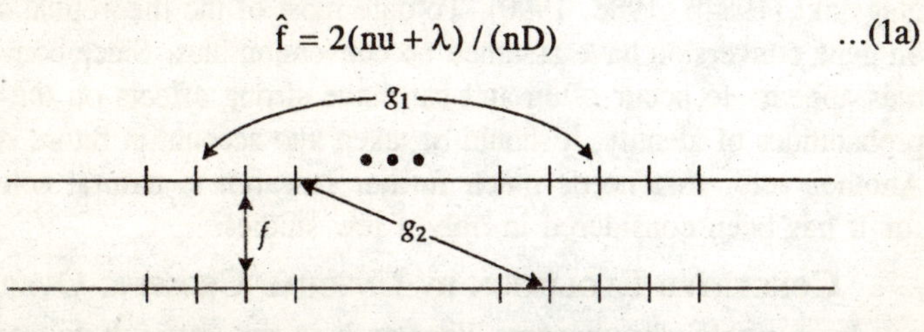

Fig. 6.9. Three identity probabilities.

$$\hat{g}_1 = \hat{g}_2 = 2\lambda / (nD) \qquad \text{...(1b)}$$

where

$$D = 2u + 2\lambda / n + 8Nu(u + \lambda) \qquad \text{...(1c)}$$

and l is the conversion rate per repeat per generation. Note that $\hat{g}_1$ and $\hat{g}_2$ are equal and that all three identity probabilities are independent of the crossover frequency.

It is assumed that each allele mutates to a new allele at the rate of 10^{-6} per generation, i.e., $u = 10^{-6}$. The case of no gene conversion (i.e., $\lambda = 0$) is given for comparison; in this case $\hat{g}_1 = \hat{g}_2 = 0$ because mutations at different loci are assumed to be different. Gene conversion tends to homogenize repeated genes, and so the interlocus identities $\hat{g}_1$ and $\hat{g}_2$ increase with λ, the conversion rate. Indeed, if λ is much larger than nu, then $\hat{g}_1$ and $\hat{g}_2$ can be almost as large as $\hat{f}$, the intralocus identity. For example, for $n=10$, $N=10^3$, and $\lambda=10^3$, we find $\hat{g}_1=\hat{g}_2=0.95$ and $\hat{f}=0.96$. However, if N and n are large $\hat{f}$, $\hat{g}_1$, and $\hat{g}_2$ can become quite small; e.g., for $N=10^5$, $n=50$ and $\lambda=10^{-3}$, $\hat{g}_1=\hat{g}_2=0.047$ and $\hat{f}=0.05$. Thus, if the population size or the gene family size is large and if the conversion rate is low, it is more difficult to maintain high sequence homogeneity in a family. Note that gene conversion reduces the intralocus identity because it introduces different alleles into a locus from other loci. Thus, $\hat{f}$ decreases as λ increase.

Nagylaki (1984a) has studied also the time of convergence of equilibrium (T). He showed that T is greater than both n/λ and 4nN. If the conversion rate λ is much less than $1/(4N)$, then T is very close to n/λ, i.e., T is proportional to the family size n and inversely proportional to the conversion rate λ. If λ is much greater than $1/(4N)$, then T is very close to 4nN, i.e., T is proportional to both the family size and the effective population size.

For intrachromosomal gene conversion and other case, readers may consult Ohta (1982, 1983a,b, 1984), Ohta and Dover (1983), and Nagylaki (1984b, 1988, 1990). To date most of the theoretical studies on gene conversion have assumed no conversion bias. Since conversion bias appears to occur often and can have strong effects on the above probabilities of identity, it should be taken into account in future studies. Another actor that needs much further research is natural selection, for it has been considered in only a few studies.

Concerted Evolution by Unequal Crossing-Over

In a single chromosome lineage it is not difficult to see how unequal crossing-over can homogenize repeats in a tandem array. For example, a result of an unequal crossover, each of the daughter chromosomes has become more homogeneous than the parental chromosomes. If this process is continued, the numbers of wild-type and mutant repeats on a chromosome will fluctuate with time, and eventually one type of repeat will become dominant in the family. This, in fact, has been verified by both computer simulation and mathematical analysis. Black and Gibson (1974) used a model in which the shift of genes of unequal matching at a crossover is always one gene, and duplication and deletion occur by equal chance of 1/2. Their simulation showed that the number of unequal crossover required for the loss of gene lineages increases with the family size. Perelson and Bell (1977) treated this model by a birth-death process and showed that the "crossover fixation time," which is the expected number of unequal crossover needed until a particular type of repeat becomes fixed in the family, is approximately $(2n_0(n_0-1)$, where n_0 is the initial number of repeats in the family. This result agree well with that of Ohta (1976), who used the diffusion theory.

Smith's (1974) simulation used a more realistic model in which the shift of unequal matching at the crossover is not restricted to one repeat but to a certain range (latitude), e.g., the family size may vary randomly in the range $n_0-n_0/10 \sim n_0+n_0/10$. He found that the crossover fixation time depends not only on the family size but also on the allowed latitude of shift, i.e., the larger the latitude, the shorter the crossover fixation time. For example, if the average family size is 100, the crossover fixation is faster when the allowed latitude is $95 \sim 105$ than when the latitude is $99 \sim 101$. However, the former is not different from the case of $75 \sim 125$, suggesting that beyond a certain value, increasing the allowed latitude becomes ineffective in accelerating the crossover fixation.

When the effects of random genetic drift and mutation are taken into account, the situation becomes much more complicated. Unlike gene conversion, unequal crossing-over creates shifts in the position of repeats and also fluctuations in the family size. The effect of shift of position is difficult to treat mathematically. Moreover, f is now difficult to define as it refers to the probability of identity of two genes at the same locus in two different chromosomes, and the definition of g_2 is ambiguous because it partly includes f. For these reasons, unequal crossing-over is mathematically less well studied than gene conversion. Nevertheless, in a series of studies Ohta (e.g., 1978, 1980, 1982, 1983a) has obtained approximate results for g_1 and g_2.

RELATIVE ROLES OF GENE CONVERSION AND UNEQUAL CROSSING-OVER

As a mechanism of concerted evolution, gene conversion appears to have several advantages over unequal crossing-over. First, gene conversion causes no change in gene number, whereas unequal crossing-over changes the number of repeated genes in a family. Such a change may cause dosage imbalance. For example, deletion of one of the two duplicated α-globin genes in man gives rise to the α-thalassemia-2 syndrome. Second, gene conversion can act as a correction mechanism not only on tandem repeats but also on dispersed repeats. In contrast, unequal crossing-over has certain restrictions when repeats are dispersed. It probably can act effectively on repeats dispersed on nonhomologous chromosomes if they are located at the ends of chromosome arms as in the case of 5S RNA genes in *Xenopus* and that of rRNA genes in man and apes. But, it will be greatly restricted if the dispersed repeats are located inside chromosomes, as in the case of rRNA genes in mice and in *Drosophila melanogaster*. If repeats are dispersed on a chromosome, unequal crossing-over can cause deletion or duplication of the genes between the repeats. Fig. 6.10 shows a hypothetical case of unequal crossing-over between two repeat clusters, resulting in the deletion of a unique gene in one chromosome and a corresponding duplication in the other chromosome. Third, gene conversion can have a preferred direction, and a small disparity can have a large effect on the probability of fixation of mutant repeats.

In large families of tandemly repeated sequences, however, unequal crossing-over may be equally or even more important than gene conversion for concerted evolution. First, in such families, the number of repeats can apparently fluctuate greatly without causing significant adverse effects. This is suggested by the observations that the number of rRNA genes in *Drosophila* varies widely among individuals in the

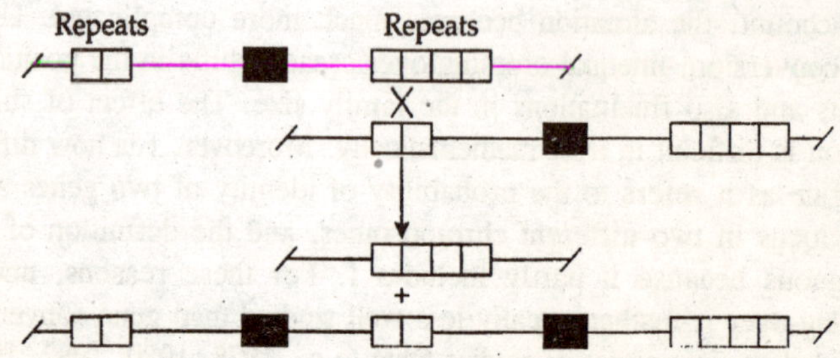

Fig. 6.10. Crossing-over involving dispersed repeats.

same species and that the number of 5S RNA genes is about 24,000 copies in *Xenopus laevis* but only 9000 copies or so in *X. borealis*. Moreover, in humans, several families of tandem repeats that exhibit extraordinary degrees of variation have been found. Second, as suggested by the data, unequal crossing-over probably occurs more often than nonallelic gene conversion. Third, a gene conversion event may involve only a small region, say, from a few bases to several hundred bases. In the examples of gene conversion in the γ-globin genes proposed by Slightom et al. (1980) and Scott et al. (1984), the DNA segments involved are all relatively short. By contrast, the number of repeats exchanged in an unequal crossing-over event can be large. In Szostak and Wu's (1980) study with yeast rRNA genes, a single unequal crossing-over event affects on average seven repeats. Obviously, the larger the number of repeats exchanged in an unequal crossing-over event is, the higher the rate of concerted evolution will be. This advantage of unequal crossing-over may be enough to offset the advantages of gene conversion. As discussed on earlier, unequal crossing-over seems to have played a more important role than gene conversion in the concerted evolution of the "360" and "500" repeat sequence families in *Drosophila*. These repeated sequences do not seem to have any function and so unequal crossing-over probably can occur freely within each repeat cluster. Unequal crossing-over has also been suggested to have played a much more important role than gene conversion in the concerted evolution of immunoglobulin V_H gene family in mouse. Of course, the observed lower rate of gene conversion in this family might have been due to detection bias, for it is generally much easier to detect unequal crossing-over than gene conversion.

Other Mechanisms of Concerted Evolution

In addition to gene conversion and unequal crossing-over, several other mechanisms have also been proposed for concerted evolution. These include the master-slave hypothesis, the saltatory replication hypothesis, replication-slippage or slipped-strand mispairing, and duplicative transposition.

Like unequal crossing-over, slipped-strand mispairing (SSM) is an expansion-contraction process that can lead to homogenization of a tandem repeat family. As noted, SSM is probably more important than unequal crossing-over in the generation of tandem arrays of short repeats. Recent molecular data obtained from such techniques as Southern blot analysis and the polymerase chain reaction method have revealed the existence of numerous microsattelite DNAs in man. SSM seems to have played a major role in the sequence homogenization of such loci. Of course, SSM and unequal crossing-over can operate on the same gene family. Unlike the prokaryotic large subunit (LSU) rRNA genes, the eukaryotic LSU rRNA genes each consist of not only core regions but also expansion segments. While unequal crossing-over is probably mainly responsible for the coevolution of the whole rRNA repeats, SSM seems to be important for the coevolution of the expansion segments within a repeat.

The master-slave hypothesis postulates that in each generation new slave copies are replicated from the master sequences; thus, variant repeats are removed in each generation. This hypothesis is incompatible with the observation that neighbouring repeats in the rRNA gene family in *Xenopus* can be unequal in length.

The saltatory-replication hypothesis proposes that a recent amplification of a sequence is the basis for intraspecific homogeneity. Under this hypothesis, the homogeneity of a family should slowly disappear, owing to accumulation of mutations, and for two species that diverged after the sequence amplification the intraspecific variation should be approximately equal to the interspecific variation. As discussed above, this prediction is not supported by observation.

Duplicative transposition involves the duplication and movement of a sequence from one position to another. A sequence with a high propensity to duplicate with movement of one copy to another location has a good chance of spreading in a genome, and hence spreading through a sexually reproducing population. There are many such families in the genome of higher vertebrates. However, if the high homogeneity among the members of a family is due to recent expansion rather than being maintained by exchange of information among members, then the family is not really undergoing a *concerted* evolution, as there is little interaction between members.

Factors Affecting the Rate of Concerted Evolution

In the above sections we discussed the effects of family size and population size on the rate of concerted evolution of a multigene family. In the following sections we shall discuss other factors.

Arrangement of Repeats

How cohesively the members of a repeated sequence family can evolve together depends on how they are arranged in the genome. There are, roughly speaking, two types of arrangement. First, the members are highly dispersed all over the genome. For example, *Alu* repeats are interspersed with single-copy sequences throughout the genome. This type of arrangement is least favourable for concerted evolution because it greatly reduces the chance of unequal crossing-over and gene conversion and because unequal crossing-over would often lead to disastrous cytological consequences. The high similarities among *Alu* repeats is largely due to recent saltatory amplification of the family rather than concerted evolution.

Second, all members of a family are clustered either in a single tandem array or in a small number of tandem arrays. As mentioned above, the rDNA family in *Xenopus* is present in a tandem array. This arrangement is most favourable for unequal crossing-over and gene conversion to operate. If the repeats are located on more than one chromosome, the rate of unequal crossing-over would be greatly reduced, unless all the clusters are at the ends of chromosome arms, as in the case of the rDNA family in humans. Moreover, the rate of gene conversion might also be reduced to some extent.

Functional Requirement

In terms of functional requirement, there are two extreme types of multigene families. One is that the function has an extremely stringent structural requirement, and often requires a large amount of the same gene product. The rRNA genes and the histone genes are well-known example. The other extreme is that the function requires a large amount of diversity. The immunoglobulin genes and the major histocompatibility genes belong to this category. If all other conditions are the same for the two different types of gene families, the rate of concerted evolution is expected to be higher in the former type than in the latter type of families. Indeed, according to Gojobori and Nei's (1984) estimate, the rate is roughly 100 times higher in human rDNA family than in the mouse immunoglobulin V_H family. This large difference in rate seems to be mainly due to the difference in selection scheme between the two families. In rRNA genes, homogeneity appears to be advantageous, and since unequal crossing-over and gene conversion both have the effect of reducing the genetic variability among multigenes, they apparently operate effectively for maintaining the homogeneity or rRNAs. In the case of V_H genes, however, an individual who has many copies of the same DNA sequence owing to unequal crossing-over or gene conversion seems to be at a disadvantage and may thus be eliminated from the population. We note further than homogeneity, on the one hand, and unequal crossing-over and gene conversion, on the other hand, would go in a cycle. Unequal crossing-over and (nonallelic) gene conversion both arise as a consequence of sequence misalignment between nonallelic repeats; there is indeed evidence from fusion gens in the human β-globin complex that the rate of unequal crossing-over increases with the length of sequence identity between duplicate genes. Thus, the higher the homogeneity among the repeats in a family is, the higher the rates of unequal crossing-over and gene conversion will be (and vice versa). For this reason, the rates of unequal crossing-over and gene conversion in the rDNA family would be higher than those in the V_H family. This could be an important reason for the large difference in the rate of concerted evolution between the two families. It goes without saying that, in a multigene family whose stringency of functional requirements is intermediate between the above two extremes, the condition for concerted evolution would be moderately favourable.

Structure of Repeat Unit

By the structure of repeat unit we mean the numbers and sizes of noncoding regions (e.g., introns and spacer regions) and coding regions contained in a repeat unit. Since noncoding regions generally evolve rapidly, it would be difficult to maintain high homology among the repeats in a family, if each repeat contains large or many noncoding regions. As just mentioned, the rates of unequal crossing-over and gene conversion are expected to decrease with decreasing intraspecific homogeneity. Thus, the structure of repeat unit should play a role in concerted evolution.

Zimmer et al. (1980) estimated that in higher primates the rate of concerted evolution in the α-globin gene region is 50 times faster than that in the β-globin gene region. They suggested that the rate in the β-globin region has been greatly reduced because the introns and flanking sequences are highly divergent between the two β genes. It is interesting to note that the β genes have introns that are several times longer than those in the α genes and that the intergenic region between the two β genes is 2400 bases longer than that between the two α genes. Zimmer et al. (1980) suggested that the larger introns and intergenic region in β genes arose as a response to selection for reduction in the rate of unequal crossing-over, which produces a single gene state in which the major gene is under the control of the minor promoter. Because the noncoding sequences in the β region are highly divergent, they can act as barriers to unequal crossing-over. As the length of such sequences goes up, their effectiveness as barriers will arise. Note, however, that the selective advantage for a reduction in the rate of unequal crossing-over would be very small, at most as large as the amount of reduction in rate. Thus, the larger introns and intergenic region might have arisen by chance rather than by selection. It is also possible that the introns and the intergenic region were already large before the divergence of the higher primates, and this has promoted the divergence between the two β genes, including divergence in regulation. On the other hand, the introns and intergenic region in the α genes could have been small or could have been reduced since the divergence of higher primates, and as a consequence, concerted evolution has occurred at a higher rate in the α genes than in the β genes.

Natural Selection and Biased Gene Conversion

Positive natural selection can accelerate the process of concerted evolution because the rate and probability of fixation of a variant repeat that is favoured by natural selection will be larger than those for a selectively neutral variant repeat. In fact, since unequal crossing-over and gene conversion can cause a large variation in the number of variant repeats per genome among individuals in the population, natural selection can become very strong among individuals in the population. By contrast, negative natural selection can effectively eliminate deleterious new variants. Therefore, positive selection can accelerate the differentiation between repeat members from different species whereas negative selection can maintain high homogeneity among repeat members within a species.

The effect of biased gene conversion on the evolution of multigene families would be similar to that of natural selection, though it may be weaker. The probability and rate of fixation of a new variant repeat is increased, if it is favoured by gene conversion. On the other hand, if the predominant type of repeat is favoured by gene conversion, then new variants may be eliminated quickly from the genome. Thus, like natural selection, biased gene conversion can accelerate between-species differentiation and maintain high within-species homogeneity.

Both natural selection and biased gene conversion work more effectively in large populations than in small ones because the effect of random genetic drift decreases with increasing population size.

Evolutionary Implications

Since concerted evolution of members of a gene family is different from the traditional notion of independent evolution of each family members, it has some profound evolutionary implications.

Spread of Advantageous Mutations

Gene conversion, unequal crossing-over, and other mechanisms of concerted evolution allow spreading of a variant repeat to all family members. This capability of horizontal spreading has profound evolutionary consequences, for an advantageous mutant repeat can replace all other repeats and becomes fixed in the family. We note that the advantage that a single variant can confer to an organism is usually very limited. The advantage would, however, be greatly amplified if the mutation spreads to many or all members in the family. Thus, through concerted evolution, a small selective advantage can become a great advantage. In this respect, concerted evolution surpasses independent evolution of individual gene family members.

Arnheim (1983) has compared the evolution of RNA polymerase I transcriptional control signals with that of RNA polymerase II transcriptional control signals. The former RNA polymerase transcribes only the rRNA genes whereas the latter transcribes all protein-coding genes. RNA polymerase I transcription control signals with that of RNA polymerase II transcriptional control signals. The former RNA polymerase transcribes only the rRNA genes whereas the latter transcribes all protein-coding genes. RNA polymerase I transcription control signals appear to have evolved much faster than the signals for RNA polymerase II. For example, in cell-free transcription systems, mouse rDNA clone does not work in human cell extract, but clones of protein-coding genes from astonishingly diverse species are transcribed in heterologous systems (e.g., silk worm genes in human cell extracts and mammalian genes in yeast).

Arnheim (1983) argues that in the case of RNA polymerase I transcription units, mutations that favourably affect transcription initiation could be propagated throughout the rDNA multigene family as a consequence of concerted evolution. It has also been suggested that even mutations that disfavourably affect transcription initiation to some extent can spread through the rDNA family; since such mutations initially affect only a small number of rDNA genes and so would be subject to weak selection, they may persist long enough to allow mutations in the rDNA genes. Such correlated changes between the rDNA genes and the RNA polymerase I gene are a type of molecular coevolution. In contrast, in the case of RNA polymerase II transcription units, advantageous mutations affecting transcription initiation that occur in any gene would not be expected to be propagated throughout all genes, for they belong to many different families.

Retardation of Divergence of Duplicate Genes

The traditional view of the creation of a new gene has been that a gene duplication occurs and one of the two resultant genes gradually diverges and becomes a new gene. It is now clear that the process may not be so simple as previously assumed. As long as the degree of divergence between the two genes is not large, the diverged copy may be deleted by unequal crossing-over or converted to the conserved copy by gene conversion. In the former case it requires another duplication to create a redundant copy, while in the latter case divergence must start again from scratch. Thus, divergence of duplicate genes may produced much more slowly than traditionally thought, and for this reason the chance of creation of a new gene from a redundant copy is reduced. On the other hand, gene conversion may also prevent a redundant copy from becoming nonfunctional for a long period time or, alternatively, may enable a "dead gene" (pseudogene) to be "resurrected".

Evolutionary History of Duplicate Genes

It has been customary to assume that, following a gene duplication, the two resultant genes will diverge monotonically with time. Under this assumption it is rather simple to infer the time of the duplication event. For example, the protein sequences of human β and δ globins are more similar to each other than to rabbit β1, or mouse β major and minor sequences. It has therefore been inferred that the two human genes were derived from a duplication event about 40 million years (Myr) ago, long after the mammalian radiation (about 80 Myr ago). The evolutionary history of duplicate genes inferred in this traditional manner may be erroneous in view of the fact that duplicate gene can correct each other.

In fact, it has recently been suggested that the β and δ genes originated from a duplication event before the mammalian radiation. This suggestion is based on the observations that the large intron and the 3' untranslated region of rabbit pseudogene ψβ2 are more similar to human δ than to rabbit β1, and that the pseudogene ψβ3 in mouse is similar to δ at its 3' end. This hypothesis is not supported by the observation that no obvious descendant of δ is found in the goat β-globin family, but if it turns out to be true, it will provide an excellent example how gene correction events can obscure the evolutionary history of duplicate genes. In a large multigene family, gene correction events will occur more often, and it will be even more difficult to trace the evolutionary relationships of the family members.

Generation of Gene Variation

From the evolutionary point of view there is an analogy between evolution of multigene families and evolution of subdivided populations. One may regard each repeat in a multigene family as a deme in a subdivided population. Then, transfer of information between repeats would be equivalent to migration of genes or individuals between demes. It is well known that migration will reduce the genetic difference between demes but will increase the amount of genic variation, e.g., the number of alleles, in a deme. Similarly, transfer of information between repeats will reduce the genetic difference between repeats but will increase the amount of genic variation at a locus. Note that the intralocus variation, 1–f, is lower in the absence of gene conversion, i.e., $\lambda=0$, than in the presence of gene conversion. Note also that $1-g_1$ and $1-g_2$, the interlocus variations, are 1 in the absence of gene conversion but become lower in the presence of gene conversion.

It is well known that some major histocompatibility complex (MHC) loci are highly polymorphic in humans, mice, and certain other mammals. For example, allelic gene products at a H-2 antigen locus from different inbred mouse strains can differ extensively in amino acid sequences and hence in serological properties. More than 50 different alleles at each of the H-2 K and H-2 D loci have been observed in mice. Weiss et al. (1983) and many other have suggested that the large numbers of alleles at the H-2 loci are mainly generated by gene conversion-like events; donor sequences could be from the H-2 loci and from nearby loci. An alternative explanation is that these alleles have persisted in the population for a long time, probably being maintained by overdominant selection. These are two extreme views and it is possible that both gene conversion and overdominant selection have been responsible for the high genetic variation at MHC loci.

7

EVOLUTIONARY RECONSTRUCTION

A knowledge of the phylogenetic relations among species is essential for many other inferences in biology, and consequently a major effort has been made to reconstruct the tree of life. The fundamental principle used in most phylogenetic inference is the principle of parsimony, and a widely used version of parsimony requires a distinction between "unrooted" and "rooted" trees.

A *phylogenetic tree*, or *phylogeny*, or *tree*, for a group of species is a branching diagram that shows, for each species (or group of species), with which other species (or group of species) it shares its most recent common ancestor. A phylogeny implicitly has a time axis, and time usually goes up the page. In the figure, species *A* and *B* share a more recent common ancestor with one another than either shares with any other species (or group of species). This chapter discusses how phylogenetic relations are inferred from imperfect evidence. They must be inferred because we exist now, whereas the splitting events and common ancestors existed in the past and cannot be directly observed.

How do we know, for example, that humans and chimpanzees share a more recent common ancestor than either shares with the amoeba? The quick answer is that humans and chimpanzees look more similar. For phylogenetic inference, the vague expression "look more similar" must be made more precise. We can divide up the phenotypic appearance of an organism into a series of characters and character states. We can then say that humans and chimpanzees have similar appearances because they share many more characters in common than either does with the amoeba. Humans and chimpanzees share vertebrate

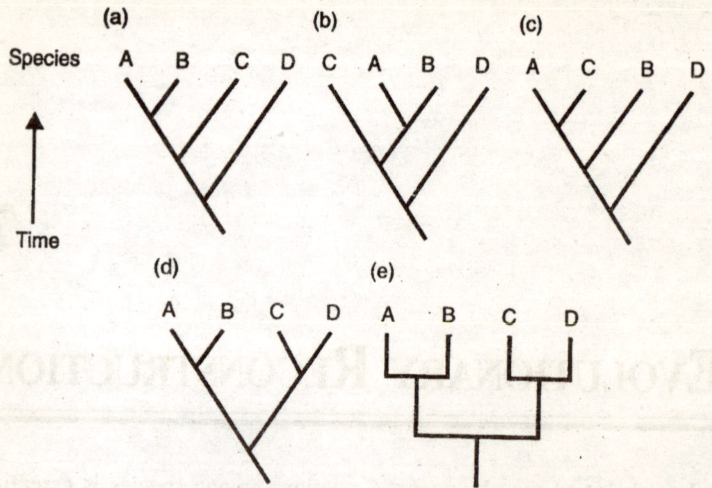

Fig. 7.1. A phylogeny shows, for a group of species, the order in which they share common ancestors with one another.

characters such as brains and backbones, mammalian characters such as lactation, and great ape characters such as their distinctive molar teeth and absence of a tail. Humans and amoebas do have characters in common—many of the characters of cellular metabolism, for example—but these characters are equally shared with chimpanzees and do not suggest a separate grouping of human and amoeba.

A more fundamental principle underlies phylogenetic reconstruction than that of shared characters, however. A phylogeny is more plausible if it requires less, rather than more, evolutionary changes in character states. This is the principle of *parsimony*. Phylogenetic inference using parsimony proceeds in two stages, and to understand them we need to distinguish between a *rooted* and an *unrooted tree*. Full phylogenies are "rooted" trees—that is, the position of the deepest ancestor, or "root" within the tree is specified. An unrooted tree shows the branching relations between the species but does not show the position of the deepest common ancestor. Thus, an unrooted tree consists of a phylogenetic tree with the time dimension removed. It can be rotated through any angle on the page without altering the information in it, because the only information it contains is the branching topology of the species. An unrooted tree is, therefore, less informative than a rooted tree.

A rooted tree is compatible with only one unrooted tree. In contrast, an unrooted tree is compatible with a number of possible

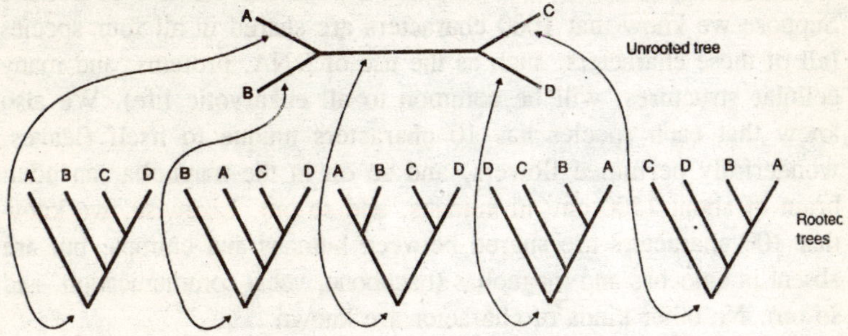

Fig. 7.2. Unrooted and rooted trees.

rooted trees (of which at most one will be the true one), depending on where the common ancestor, or root, lies. The smallest number of species for which a nontrivial unrooted tree can be drawn is four. A group of three (or two, or one) species can have only one possible unrooted tree, and it is therefore trivial to write this tree down. Four species have three possible unrooted trees, however, and it is nontrivial to assert that the true unrooted tree is any one of those alternatives.

The two stages in the method of parsimony are first to infer the unrooted tree for a set of species, and then to locate its root. How can we infer the unrooted tree? Let us consider the simplest case, in which we have four species. We can add one species, such as a magnolia tree, to the human, chimpanzee, and amoeba in the example above. The procedure is as follows. First, write out all possible unrooted trees for the species. In this case, and throughout the rest of the chapter, we shall confine ourselves to the case of bifurcating (or dichotomous) trees, so three trees are possible for the group of four species. (Phylogenetic research normally confines itself to trees with bifurcating branches, because most real trees probably have bifurcating, rather than multiple, branches.) Next, given a knowledge of the character states in each species, count the minimum number of evolutionary events (that is, changes in character states) implied by each tree. The best estimate of the true unrooted tree is the one requiring the least evolutionary change.

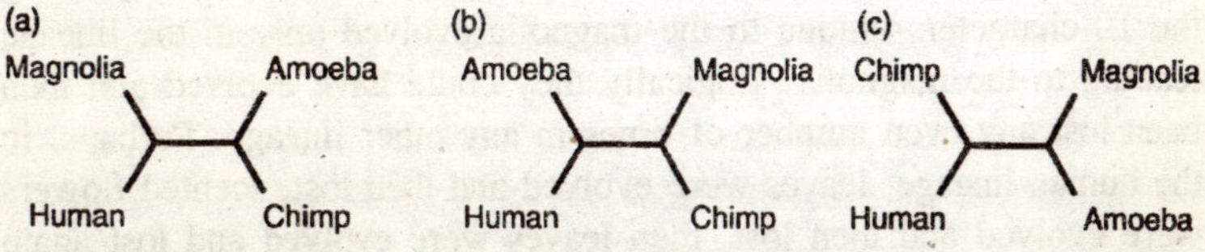

Fig. 7.3. Possible phylogenetic relations of amoeba, chimpanzee, human, and magnolia.

Which of the three trees requires the smallest number of events? Suppose we know that 1000 characters are shared in all four species (all of these characters, such as the use of DNA, proteins, and many cellular structures, will be common to all eukaryotic life). We also know that each species has 10 characters unique to itself (leaves, wonderfully perfumed flowers, and so on in the magnolia, an adult brain of about 1500 cm^3 in humans, and so on). Likewise, we know that 100 characters are shared between humans and chimps, but are absent in amoebas and magnolias (backbone, vocal communication, and so on). No other kinds of character are known.

We can now count the smallest number of evolutionary events implied by each unrooted tree. The 1000 common characters could have evolved once each in the common ancestor of all four species and been retained throughout the phylogeny. That possibility would require a total of 1000 evolutionary events (one evolutionary origination for each character). Each species' 10 unique characters can have evolved in the lineage leading to that species, making 40 evolutionary events in all. The 100 characters shared by humans and chimps are more puzzling. The smallest number of events producing the observed distribution of characters is 200. They could have evolved separately in the human lineage and in the chimp lineage (100 events in each, making 200). Alternatively, if the branch leading to the magnolia contains the common ancestor of the group of four species, all 100 could have evolved once between that common ancestor and the common ancestor of the amoeba, human, and chimp. They would then need to be lost in the lineage leading to the amoeba, making 100 gains and 100 losses, or 200 events in all. (The same argument can be made if the common ancestor of the four species appears in the branch leading to the amoeba; the characters then must be lost in the lineage leading to the magnolia.) We do not need to know the real pattern of character changes to count the number of events implied by the tree, because either way the total number—and therefore the minimum number—of events is 1240.

The tree is logically compatible with any number of events between 1240 and infinity. The minimum number was obtained by supposing the 10 characters unique to the magnolia evolved only in the lineage leading to the magnolia. Logically they could have evolved and then been lost any even number of times in any other lineage. Perhaps, in the human lineage, leaves were evolved and then lost, scented flowers were evolved and then lost, then leaves were evolved and lost again

... and so on to infinity. The interesting number is the minimum number of events implied by the tree, not the number of events that logically could have occurred.

The same calculation also implies 1240 events. However, the 100 characters shared by humans and chimps need to have evolved only once, in the common ancestor of humans and chimps, and would not have to be lost again. Thus, this tree only requires 1140 events. It is the most parsimonious unrooted tree, and (according to the principle of parsimony) represents the best estimate of the real unrooted tree. The unrooted tree completes stage one of the inference. The second stage is to locate its root, to turn it into a phylogeny.

The example illustrates another important distinction, between informative and uninformative characters (or evidence). An uninformative character, in phylogenetic inference, is one that is equally compatible with all possible unrooted trees for the species under analysis. The 1000 characters shared by all four species are uninformative. They require the same number of events in (a), (b), and (c) and, therefore, do not help us calculate which tree is more likely to be correct. The 4×10 unique characters are also uninformative because they require the same 40 events in all three unrooted trees. All of the difference in the total number of events required by the unrooted trees is contributed by the characters shared between chimpanzees and humans. Thus, these characters are the only informative. They require fewer events in (c) than in (a) or (b) and allow the inference that (c) is most likely.

With this knowledge, we could have performed the same analysis in another way. We could first have examined all 1140 characters for which we have evidence and divided them into informative and uninformative characters. Then we could discard all the uninformative characters, and count the number of events implied by the unrooted trees for the informative characters alone. The result would be 200 events in (a) and (b) and 100 for (c). Using this method, again represents the most parsimonious tree. The result, and the underlying logic of the argument, is the same in the two methods. But the second method, in which all uninformative evidence is identified and discarded before the events are counted can often be quicker and clearer.

Parsimony Principle

How can we justify the parsimony principle? Why is a phylogeny requiring less evolutionary events a more plausible inference than one requiring more? The parsimony principle is reasonable because

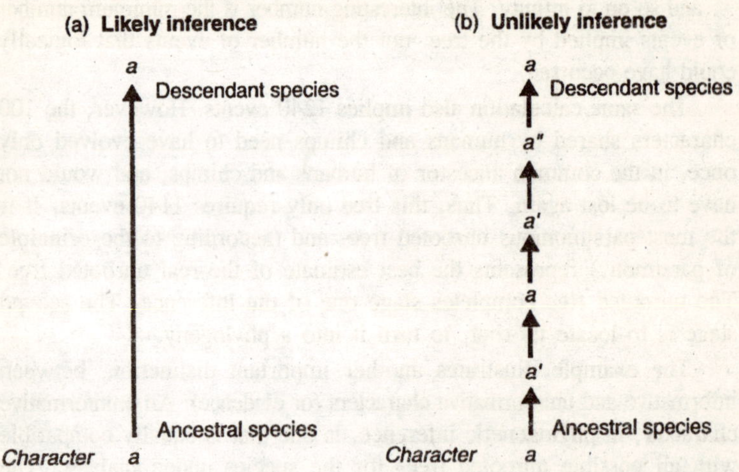

Fig. 7.4. The same character is found in both a descendant species and one of its ancestors.

evolutionary change is improbable. Suppose we know that a modern species and one of its ancestors both have the same character state. Parsimony suggests that all intermediate stages in the continuous lineage between the ancestor and modern species possessed that same character state. As we have seen, a large number of changes—indeed, an infinite number—could logically have occurred between ancestor and descendant. A change followed by a reversal of that change is an unlikely occurrence, however. Each change requires a gene (or set of genes) to arise by mutation and then to be substituted, either by drift (if the change is neutral) or by selection; both these processes are improbable. It is much more likely that the same character would have been continuously passed on, in much the same form, from ancestor to descendant by simple inheritance. We know this process is plausible because it happens every time a parent produces an offspring: the parental characters are passed on.

For the characters shared between humans and chimps, this argument is particularly powerful. Chimps and humans share whole complex organ systems like hearts, lungs, eyes, brains, and spinal cords. The initial evolution of each of these characters required improbable mutations, and natural selection operating over millions of generations. It is evolutionarily improbable to the point of near impossibility that the same changes would have evolved independently in the two lineages after their common ancestor. By contrast, there is nothing improbable about postulating that the characters could have been passed on via passive inheritance from the common ancestor of chimps and humans to the modern descendants.

For some characters other than the complex morphological characters shared between humans and chimps, this argument is less powerful. At the other extreme, if we find one nucleotide, at a particular site in the DNA, shared between two species, there is a 25% probability that it could be shared by chance, and the principle of parsimony does not strongly suggest that the nucleotide has not changed through all evolutionary intermediates between the two species.

While the argument's power varies from case to case, evolutionary change in all characters is improbable to some extent, as compared with simple inheritance. The principle of parsimony, therefore, has a sound evolutionary justification. In conclusion, it is more likely that a character will be shared by common descent than by independent, convergent evolution. For any set of species, a phylogeny requiring less evolutionary change is more plausible than one requiring more.

Parsimony and Distance Statistics

So far we have encountered two principles of phylogenetic inference, and we use these principles throughout this chapter. These principles are:

1. *Parsimony*. Species are arranged in a phylogeny such that the smallest number of evolutionary changes is required.
2. *Distance* (or *similarity*). Species are arranged in a phylogeny such that each species is grouped with the other species with which it shares the most characters.

In the case of the chimpanzee, amoeba, and human, the argument that humans and chimpanzees share a more recent common ancestor because they have more similar appearances implicitly used measurements of distance. The counts of evolutionary events in the three unrooted trees for the human, chimp, amoeba, and magnolia used the principle of parsimony.

In easy cases like that of humans, chimpanzees, and amoebas, the distance and parsimony principles give the same result, and it might seem that it does not matter which is used. In other cases, however, results differ for these principles. The parsimony principle is then more reliable, because it has a better theoretical justification. Distance statistics remain an important method in phylogenetic inference, because under certain circumstances they are almost as reliable as parsimony—and they can often be collected more rapidly. To know when we can rely upon any given method, it is necessary to understand the underlying principle of phylogenetic inference.

Phylogenetic Relationship

In the imaginary example, all of the characters implied the same phylogeny. Some characters were unique to each species, other characters were shared by all species, and still other characters were shared by only chimpanzees and humans. The 1140 characters "agreed" in the sense that we could divide them into two sets and still obtain the same answer. If, for example, we divided the 1000 + 40 + 100 characters into two sets of 500 + 20 + 50, the phylogeny inferred from each of those two sets would be the same. Now imagine that, in addition to the 100 characters shared between chimpanzees and humans (and absent from magnolias and amoebas), we have 50 characters shared by magnolias and humans (and absent from chimpanzees and amoebas). In this case, the characters do not agree. If we concentrate on the characters shared between humans and magnolias, the phylogeny is correct; on the other hand, the characters shared between humans and chimps favour the phylogeny.

Conflict between characters is typical in real phylogenetic problems. (In a case like the human, chimpanzee, amoeba, and magnolia, conflict might not present a problem, but the answer is obvious. It would be more accurate to say that the cases in which no character conflict occurs have all been worked out, and the real problems left for phylogenetic research are the cases of conflicting evidence.) The reptiles and birds are an example of such conflict in characters. Some characters, such as gait, suggest one phylogenetic grouping; other characters, such as skull anatomy, suggest another tree.

When characters conflict, how should we proceed? There are two main answers, and they are related rather than mutually exclusive.

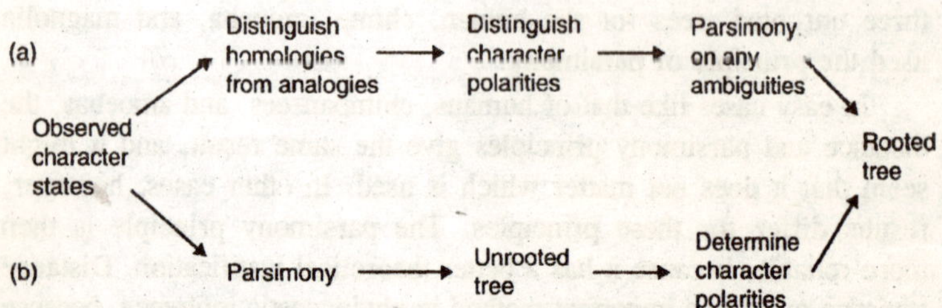

Fig. 7.5. Two routes of phylogenetic inference. (a) Two-stage character analysis to reduce conflict, followed by parsimony. (b) Proceed straight to parsimony without character analysis.

They almost form, in practice, two routes to infer a phylogeny. One possibility is to resolve the conflict directly by parsimony. If humans and chimpanzees share 100 characters, and humans and magnolias share 50 characters, the phylogeny putting the humans and chimpanzees together will be more parsimonious: that is, it will require fewer evolutionary events. As we shall see, this method is essentially the same as that used with molecular evidence.

The other possibility is to analyze the characters, to see whether some serve as more reliable phylogenetic indicators than others. This "classical" cladistic method has particularly been used with morphological evidence. The unreliable shared characters, once they have been identified, can be discarded. Any remaining conflict can then be resolved by parsimony.

We can, therefore, either go straight to the principle of parsimony, or sift the evidence first and then use the principle of parsimony. The fundamental evolutionary logic of the two methods is in many ways identical, and the different applications with morphology and molecules mainly derive from the level of biological understanding of the evidence.

Homologies and Analogies

The first stage in the top route, in which we analyze the characters to assess their phylogenetic reliability, is to distinguish homologies from analogies. We encountered this distinction when discussing classification, but there we assumed that the character analysis had been performed, and focused instead on how to use the different characters in classification. In this case, we will consider how to analyze the characters. What is the difference between homologous characters, or *homologies*, and analogous characters, or *analogies*? A homology is a character shared between two or more species that was present in their common ancestor; an analogy is a character shared between two or more species that was not present in their common ancestor. Homologous similarity reveals a phylogenetic relationship, while analogous similarity does not.

A homologous character such as the heart, or lungs, of a human and a chimpanzee is easily recognized as the same character, and is presumably derived from a common ancestor that also possessed that character. In other cases, the similarity is less obvious. The five-digit limb of tetrapods is homologous even though its form varies, and in extreme cases homologies are so subtle that it takes clever detective work to reveal them. The ear bones of mammals, for example, do not superficially resemble the gill arches of a fish. One of the classic

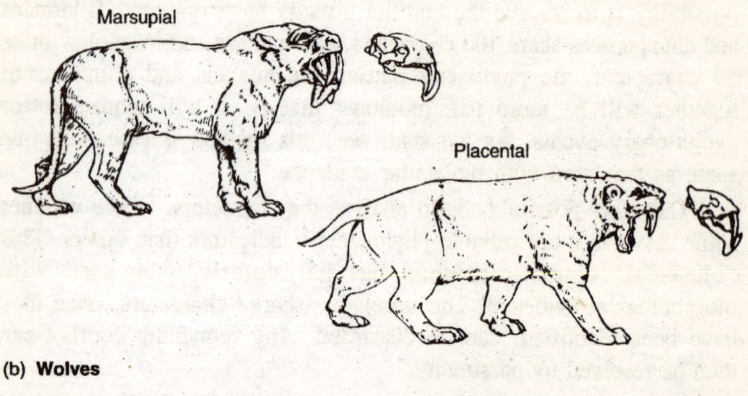

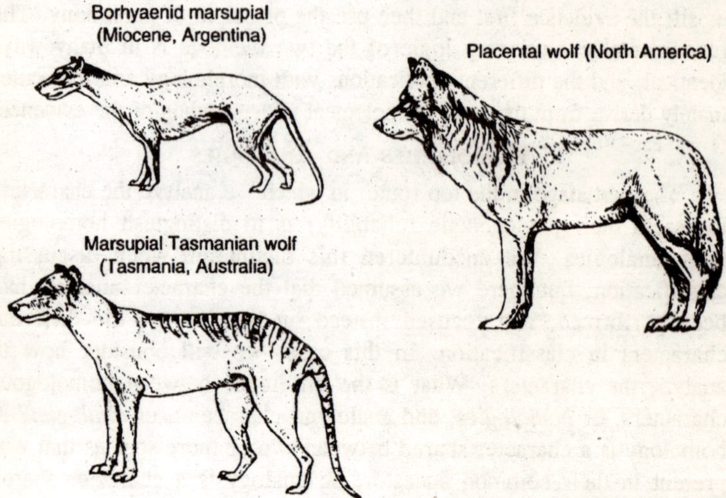

Fig. 7.6. Convergence in marsupial and placental carnivores.

discoveries of nineteenth century comparative anatomy was, however, a series of intermediates was traced between them, and these characters were found to have a common embryonic origin—that is, they are homologous. Strictly, a homologous character does not have to be identical in all species possessing it, but there does have to be shared morphological information among them.

Convergence can also be impressively deceptive. Consider a classic example of convergence from the two major groups of mammals, the marsupials and placentals. The marsupial and placental saber-toothed carnivores both evolved long, gashing canine teeth. In addition, striking similarities in skull shape and body form are found in the marsupial and placental wolves. If we inferred the phylogeny of the marsupial wolf, the placental wolf, and the kangaroo from the pattern of phenetic similarity, we should obtain the wrong answer. The two wolves are phenetically more similar, even though the marsupial wolf is phylogenetically closer to the kangaroo than to the placental wolf. The phenetic similarity between the wolves is analogous and is not due to a close phylogenetic relationship.

Thus, analogous characters do not reliably indicate phylogenetic relationships because they can fall into any grouping of species; they are not constrained to fall into phylogenetic groups. Different analogies can fall into different and conflicting groupings, and the existence of analogies is, therefore, one reason for conflict among characters. (This point forms the first part of the full argument why phenetic similarity, as measured by distance statistics, is not a reliable guide to phylogeny.)

Differences

Imagine that we have a list of 100 characters that show similarities within a group of species. The characters give conflicting indications about the true phylogeny, with perhaps 30 characters implying one phylogeny, 30 characters implying another tree, 20 still another, and

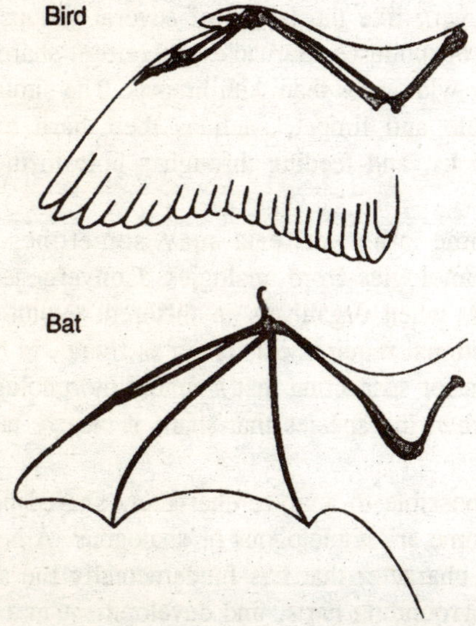

Fig. 7.7. The wings of birds and bats are analogies.

the final 20 pointing to various idiosyncratic possibilities. The task is to shorten the list and reduce its ambiguity by eliminating the analogies. How can we detect them? If a shared character is a real homology (i.e., inherited from a common ancestor), it must be the same character in the two species, and we can test it by examining the character in detail to find out whether it is the same in the different groups.

First, if a character is homologous it must have the same fundamental structure. The wings of birds and bats, for example, are superficially similar, but they are constructed from different materials and supported by different limb digits. Thus, they probably evolved independently, and not from a winged common ancestor.

Second, homologies must have the same relations to surrounding characters. Homologous bones, for example, should be connected in a similar way with their surrounding bones.

Third, the characters must have the same embryonic development in different groups. A character that looks similar in the adult forms, but develops by a different series of stages, is unlikely to be homologous. One example, which we saw earlier is the relation between a barnacle, a mollusc such as a limpet, and a crab. At least superficially, the adult form of a barnacles resembles a limpet more closely than a crab. The relations of barnacles had been uncertain for centuries until John Vaughan Thompson discovered their larvae in 1830; these forms are very like the larvae of several groups of Crustacea, and unlike those of molluscs. Barnacles, therefore, share a more recent common ancestor with crabs than with limpets. The similarities between the adult barnacle and limpet, such as their hard external armor, attachment to rocks, and feeding through a hole in the shell, are all convergent.

Finally, some other criteria may sometimes be useful in distinguishing homologies from analogies. Convergence is caused by natural selection, when organisms in different evolutionary lineages face similar functional requirements (such as flying, in birds and bats). We have grounds for suspecting that a shared morphological form may be analogous, when the species that share it clearly need it for their way of life.

Thus, it is possible to analyze characters shared between species to see whether some are homologous or analogous. A homology can be recognized as a character that has fundamentally the same structure, relations with surrounding parts, and development in a set of species. Once the homologies are (often tentatively) identified, they can be allowed into the list of evidence used to infer the phylogeny. The analogies are removed. In this way, character conflict should be reduced.

Derived Homologies

Character conflict may not, however, be reduced to zero. Of the 100 initial characters in our example, perhaps 45 could be ruled out as analogies. The 55 remaining (inferentially homologous) characters might consist of 25 characters that support one phylogeny, 15 that imply a second tree, 12 a third, and 3 that point to various other groupings. We now have a list of homologies that are shared in various combinations among the species. In most real cases, the list will contain conflict, though likely less conflict than in the list of raw, unanalyzed characters. Some of the remaining conflict may be due to mistakes—that is, some of the inferred homologies may really be analogies. However, conflict can still exist among genuine homologies because not all homologies reliably fall into phylogenetic groups.

The next stages in the analysis of the characters involve dividing the homologies into ancestral and derived homologies. An ancestral homology was present in the common ancestor of the two species, but evolved earlier and is also shared with other, more distantly related species. In contrast, a derived homology first evolved in the common ancestor of the two species and is not shared with other more distantly related species. If the two species were a human and a salmon, for example, the backbone is a derived homology; it first evolved in the common ancestor of the vertebrates—which is also the common ancestor of these two species. The presence of a distinct nucleus within all their cells is an ancestral homology; it was present in the common ancestor of the vertebrates, but evolved earlier, in the common ancestor of all eukaryotes. Thus, whether a homology is ancestral or derived depends on the set of species being compared. The backbone is a shared derived homology in a human and a salmon, but it is a shared ancestral homology in a human and a chimpanzee.

The complete analysis of a character has two stages: distinguishing analogies from homologies, followed by dividing the homologies. A character can, therefore, belong to any one of three types. The distinction matters because, of the three kinds of shared characters, only shared derived homologies constitute evidence that the two species share a more recent common ancestor with one another than with any other species under investigation. Analogies or shared ancestral homologies can be shared by species that do not form phylogenetic groups. We saw above how analogies are phylogenetically unreliable.

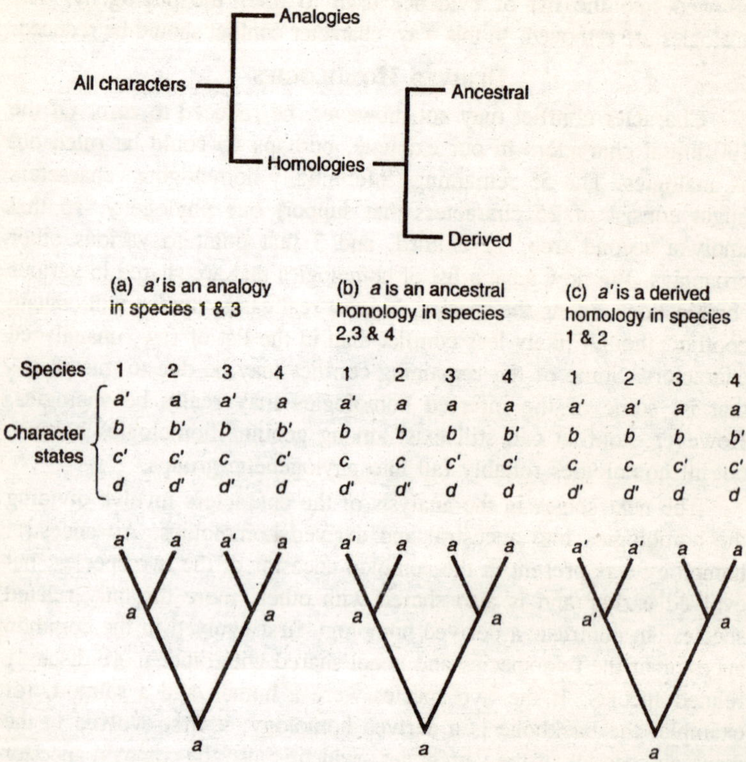

Fig. 7.8. Shared characters divide into analogies, ancestral homologies, and derived homologies. (a) a' is an analogy; (b) an ancestral homology; (c) a derived homology.

We can now complete the argument against interring phylogenies from phenetic similarity (that is, similarity with respect to all three kinds of characters together). That phenetic similarity is misleading in the case of convergence is widely appreciated. Ancestral homologies cause the same problem, however, and in a more insidious fashion. We have already seen this problem in the reptiles: a crocodile looks more like a lizard than a bird, but is phylogenetically closer to a bird than to a lizard. The crocodile and lizard share ancestral homologies—characters that were present in the common ancestor of the bird, crocodile, and lizard—that have been lost in the rapidly evolving bird lineage.

The point of these examples is not that phenetic similarity never indicates phylogenetic relationship, but that it is unreliable; in some cases, it gives the wrong answer. The unreliability of phenetic similarity gives us a reason to distrust distance statistics in phylogenetic inference, because distance statistics group species according to their phenetic similarity (exactly the same argument applies for genotypic similarity). If we had applied the informal reasoning used earlier in this chapter for the three species case of chimpanzee, human, and amoeba to the case of the lizard, crocodile, and bird, we would have blundered. Shared ancestral homologies can, like analogies, fall into conflicting groups of species.

Only derived homologies eliminate the possibility of character conflict: they must all agree on the same phylogeny. Because all characters have evolved in the same phylogenetic tree, all the shared homologies must fall into the same pattern of groups ("horizontal" transfer of characters between lineages is an exception). If the homologies and character polarities have been correctly identified in a number of characters, it is impossible for different shared derived characters to suggest incompatible phylogenies. A set of species can no more have multiple phylogenetic relations than a human family can have more than one family tree. If a human family has two family trees in its possession, at least one of them must be wrong. Likewise, if two characters suggest incompatible phylogenies, the analysis for at least one of them must be incorrect. The same is not true for analogies and ancestral homologies, however. Ten different analogies, or ten different ancestral homologies, can lead to ten different, and conflicting, groupings of species.

In summary, we have divided characters into three kinds and considered theoretically how each falls into groups of species in a phylogeny. Only shared derived homologies consistently reveal phylogenetic groups. But how can we distinguish in practice between ancestral and derived states of characters?

Polarity of Characters

The question of how to distinguish derived from ancestral homologies (also called the question of *character polarity*) has the following general form. A character has two states, which we can call *a* and *a*'. In this case, we need to know whether *a* evolved from *a*', or vice versa. In this section, we will discuss the three most important methods of discerning character polarity: outgroup comparison, the embryological criterion, and the fossil record.

Outgroup Comparison

Amniotes are the group made up of reptiles, birds, and mammals. All of these animals possess an egg membrane, called the *amnion*, during their development. It is known that amniotes are a monophyletic group—that is, they all share a unique common ancestor. Here, we can assume that the amniotes are indeed known to be a good phylogenetic group but that we do not know the relations among the different amniotes. For instance, in a set of six amniotic species (such as a mouse, a kangaroo, a bird of paradise, a robin, a crocodile, and a tortoise), does the kangaroo share a more recent common ancestor with a mouse, a bird of paradise, or one of the other animals?

Suppose we have established homologies in various characters, including reproductive physiology. The kangaroo and mouse are viviparous, and the other four species are oviparous. Did the ancestor of the group of six species breed viviparously, in which case viviparity was ancestral and oviparity derived, or did it breed oviparously, in which case evolution occurred in the opposite direction? By the method of *outgroup comparison*, the answer is found by looking at a closely related species that is known to be phylogenetically outside the group of species under investigation. The character state in that outgroup is likely to have been ancestral in the group under consideration.

In this case, we might look at a salamander, or a frog, or even a fish. They are all near relatives of the amniotes, but not amniotes themselves. These "outgroup" species almost all breed oviparously. The inference by outgroup comparison, therefore, is that oviparity is ancestral in the amniotes. Viviparity, in the kangaroo and mouse, would then be a shared derived character and oviparity, in the other four animals, would be a shared ancestral character.

In the abstract, two species, species I and 3, might share homology *a* and two others, species 2 and 4, with homology *a*'. To find whether character *a* evolved into *a*', or *a*' into *a*, we would look at a closely related species and infer that the state there is ancestral in the group of four. If the outgroup had *a*, we would infer that species 2 and 4 share a more recent common ancestor with one another than with any of the other species; the relation of species I and 3 remains uncertain (as is discussed later).

Outgroup comparison works on the assumption that evolution is parsimonious. If the character in the outgroup a is ancestral in the group of species 1-4, at least one evolutionary event must have occurred in the phylogeny: a transition from *a* to *a*' before the ancestor of species 2 and 4. If, having observed *a* in the outgroup, we had reasoned that *a*' was the ancestral state of species 1-4, at least two events must have taken place: a change from *a*' to *a* somewhere between the

(a) **Observations**

Species	1	2	3	4	Outgroup
Character state	*a*	*a'*	*a*	*a'*	*a*

(b) **Phylogenetic inference**

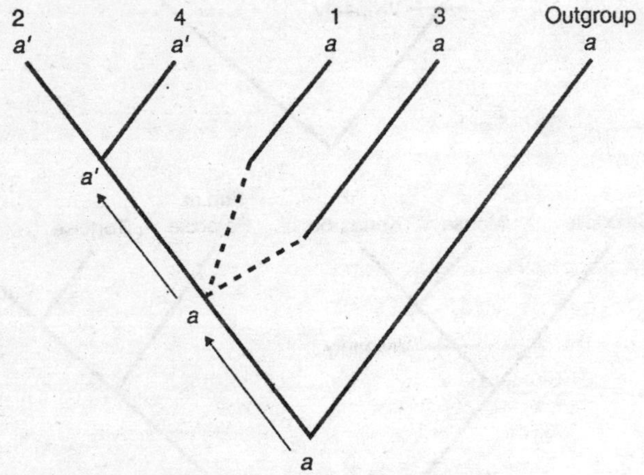

Fig. 7.9. (a) Species 1-4 have the character states given. (b) We look at a closely related species, the outgroup.

outgroup and species 1-4, and then a change from *a*' back to *a* in species 1 and 3. If the character state in the outgroup is ancestral, the fewest evolutionary events are required.

Outgroup comparison, like all techniques of phylogenetic inference, is fallible. Sometimes, one possible outgroup will suggest that one character state is ancestral, but another possible outgroup will suggest that a different character state is ancestral. The result will then depend on which outgroup is used. The method is most reliable when the closely related species that could be used as outgroups all suggest the same inference, but the method may also lead us astray in particular cases. The inference should be treated with caution, and if possible tested against other evidence.

Returning to the case of the six amniotes, the inference about viviparity provides evidence that the mouse and kangaroo share a more recent common ancestor with one another than with any of the other four animals. It provides no evidence about the relations among the

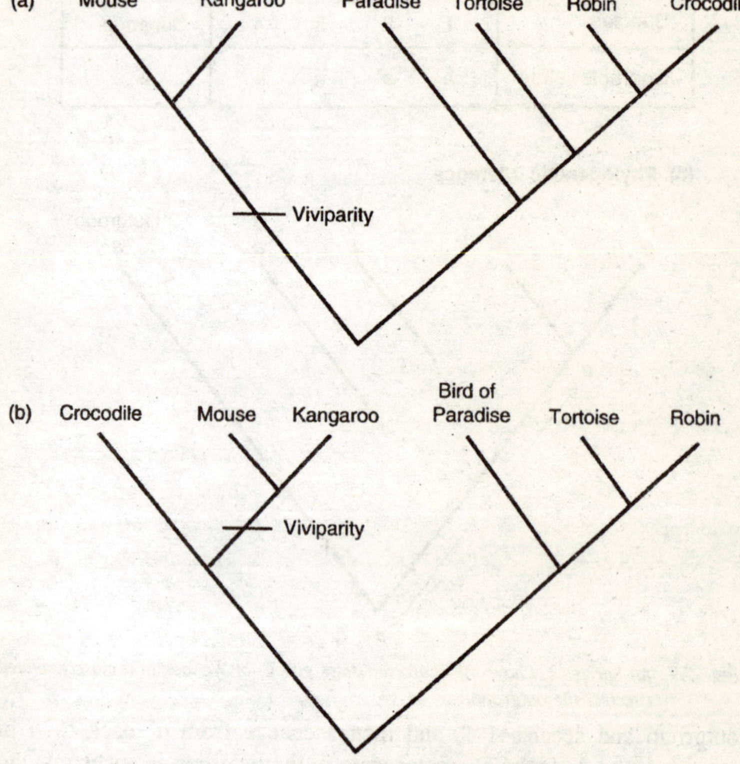

Fig. 7.10. Two possible inferences for six amniote species.

other four species. They share only an ancestral character. The evidence does not distinguish between the two types of phylogeny. The mammals could be a separate branch, or they could have branched off from any one of the other four taxa. The shared ancestral homology of oviparity does not provide evidence that the four share a more recent common ancestor with one another than with the mammals. We need more characters to sort out the four species with the ancestral character state.

Before we can use outgroup comparison, we must know something about the phylogeny. For example, we needed to know that fish and amphibians were outside the Amniota to use them as "outgroups." In practice, this is not a major problem. Outgroup comparison cannot be used when we are absolutely ignorant, but we can build on our knowledge about the phylogeny of a group (for example, that amphibians are not amniotes, but are closely related to them) to find out more (in this case, more about the phylogeny within the amniotes).

Embryological Criterion

The pre-Darwinian embryologist Karl Ernst von Baer (1792-1876) described, from microscopic observations, the course of development of several animal groups. He summarized his observations in his embryological laws. For phylogenetic inference, von Baer's first law is most important. It states: "The general features of a large group of animals appear earlier in the embryo than the special features." Cartilage, for example, is found in all fish—in cartilaginous fish such as sharks, rays, and dogfish as well as in bony fish. Cartilage is a general character; bone is a special character, being found only in bony fish. Von Baer's law predicts that, in bony fish, cartilage will appear earlier in individual development, and will transform into bone. This prediction is correct.

Von Baer's law can be given an evolutionary interpretation and then put to phylogenetic use. The characters von Baer called "general" are, in evolutionary terms, ancestral, and his "special" characters are evolutionarily derived. Thus, the successive transformations from general to special forms of a character are evolutionary changes between ancestral and derived character states. By the embryological criterion, cartilage is inferred to be an ancestral state, and bone a derived state. The bone in bony fish, therefore, evolved from a cartilaginous

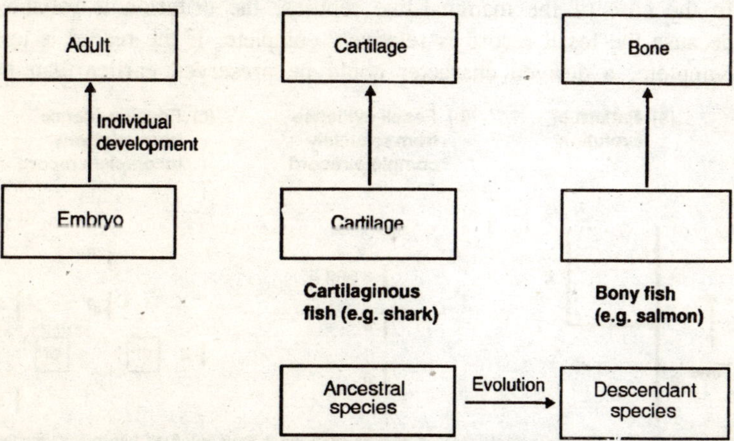

Fig. 7.11. The character that appears earlier in the development of bony fish is a general character, found in the wider phylogenetic group of bony and cartilaginous fish.

ancestry. In general, if we have a group of species and a list of homologous characters, then (if they are the kind of characters that undergo development—i.e., not things like chromosomal bands) the relative ancestral and derived character states can be inferred from their order in development.

The embryological criterion works only when von Baer's law is correct. The law's scope has never been systematically studied. It is widely accepted to have some truth, and for this reason it can be used in phylogenetic inference. It is also known to have exceptions, however. Clearly, von Baer's law should be applied only where we can be reasonably confident that it is valid, and inferences made from it should, if possible, be tested against other classes of evidence.

Fossil Record

In the evolution of the mammals from mammal-like reptiles, many characters changed over time. Posture evolved from a "sprawling" to an "upright" gait; jaw articulation and circulatory physiology also changed. Some, though not all, of these characters leave a fossil record, and we can infer which character states were ancestral and which derived by studying the earlier fossils. This type of investigation is the *paleontological criterion* of character polarity.

The reasoning could hardly be more simple. The ancestral state of a character must have preceded the derived states in fact, and therefore the earlier state in the fossil record is likely to be ancestral. In the case of the mammal-like reptiles, the criterion is reliable, because the fossil record is relatively complete. If the record is less complete, a derived character could be preserved earlier than its

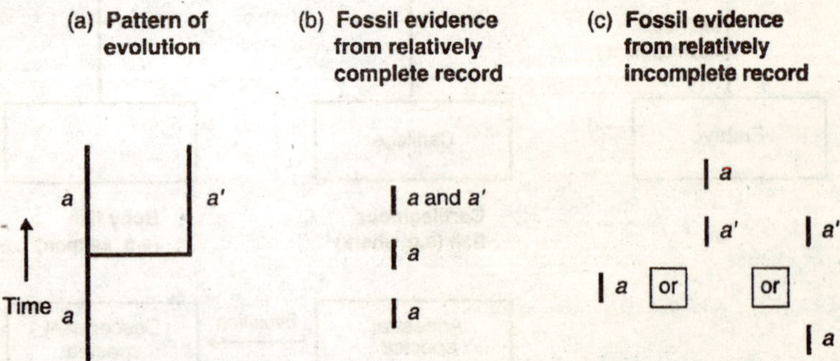

Fig. 7.12. (a) The ancestral state of a character must have evolved before its derived state. (b) If the fossil record is relatively complete, the ancestral state will be preserved in earlier fossils. (c) If the fossil record is incomplete, the derived state may or may not be preserved earlier than the ancestral.

ancestral state, and the paleontological inference will be the opposite of the truth.

For a whole fossil series like the mammal-like reptiles, we can be reasonably sure which states are ancestral. At the other extreme, of few fossils and a highly imperfect record, the evidence may be practically worthless. Most real cases lie between these extremes, and an intermediate level of confidence is appropriate.

Residual Characters

The division of shared morphological characters into analogies and homologies, and ancestral derived homologies, should reduce the conflict relative to an unanalyzed list of characters. In an ideal case, the conflict should be reduced to zero, because real derived homologies cannot conflict. The actual level of conflict, however, is likely to be reduced to something more than zero because the techniques can all yield mistakes. Convergence can be deceptively exact, and analogies can be mistaken for homologies. The criteria for determining character polarities may be inapplicable (if the character lacks a fossil record, or its embryonic development is unknown or nonexistent, or its phylogenetic relations with nearby outgroups are obscure), and even when they can be used they are still fallible. Moreover, the existence of more than one criterion may increase the uncertainty. If a character can be studied by multiple criteria, the results can be played against one another. If the criteria agree, it increases the plausibility of the conclusion; if they disagree, we face the problem of deciding which evidence to trust.

We have four options when faced with conflict in the evidence: we can scrutinize, and re-scrutinize, the contradictory results to test their reliability; we can suspend judgment; we can collect more evidence; or we can resolve the conflict by the parsimony principle (that is, pick the phylogeny that requires the fewest evolutionary events). The viability of the four will vary from problem to problem, but, after all character analysis is finished, any remaining conflict must be resolved by parsimony. Biologists vary in how much they prefer to persist with character analysis, before moving on (or resorting) to parsimony. The fact that parsimony may be used in the end does not mean the result will be the same as if it had been used initially.

Consider the example of the 100 initial characters, of which 30 pointed to phylogeny *a*, 30 to phylogeny *b*, 20 to phylogeny *c*, and 20 to other idiosyncratic arrangements. The exposure and elimination of analogies reduced the list to 25, 15, 12, and 3 (homologous) characters, respectively. The character polarities might reduce it, in turn, to 12, 3, 6, and 0 characters, respectively. The principle of parsimony is finally needed to resolve the conflict (in favour of phylogeny *a*, which is supported by 12/21 characters), but the inference becomes much more certain by virtue of the character analysis. The initial data were indifferent between phylogenies *a* and *b*, and only 30% of the data supported each of them. On the other hand, the sifted data unambiguously favoured phylogeny *a*, which was supported by almost 60% of the characters.

A final point can be made about the recognition of analogies and homologies, and of character polarities. The methods we have examined so far worked before (*prior* to) we knew the phylogeny. It is also possible to recognize the character types after (*posterior* to) the phylogeny is known. Take the case in which 12, 3, and 6 characters, as analyzed by the "prior" methods, suggest phylogenies *a*, *b*, and *c*. They cannot all be derived homologies, because real derived homologies do not conflict. Thus, some of the characters must have been misidentified. We then apply the principle of parsimony to the conflict, and deduce that phylogeny *a* is correct. The conclusion implies that the 12 characters supporting *a* are derived homologies; but the other 9 characters are not. This is the "posterior" method. The distribution of the 9 characters on phylogeny *a* will reveal, by inspection, which are convergent and which are ancestral homologies. The "posterior" method cannot be used to infer the phylogeny, because it would create a viciously circular argument. A knowledge of whether a shared character is homologous or analogous, and ancestral or derived, can be useful for other purposes, however, and a complete list of the techniques of character analysis should include both the posterior and the prior techniques.

PHYLOGENETIC INFERENCE

The sequences of proteins and DNA are both used in phylogenetic inference. Proteins blazed the trail, with the first protein to have its amino acid sequence worked out—insulin—being sequenced by Sanger in 1954. Protein sequencing became an automated process through the 1960s, and the sequences of some proteins) such as cytochrome c and hemoglobin, became available in enough species for large-scale phylogenies to be inferred. DNA sequences followed approximately 20 years later. Sanger was also the first to sequence a decent sized portion of DNA, in this case the whole genome (containing 5375 bases) of the bacteriophage ϕX174, in 1977. Since that time, DNA sequencing has expanded, almost explosively, and most current molecular phylogenetic work is concerned with DNA sequences. Many of the methods and concepts of molecular phylogenetics were established for proteins, however, and here we shall consider the two kinds of molecules together.

The deep logic of phylogenetic inference is identical for molecular and morphological characters, but the two have distinct properties and the methods and concepts used for each can appear very different. The homology/analogy distinction, in particular, differs for the two. When confronted by apparently conflicting homologies for morphological characters (like the wings of birds and bats), we first reexamine the organs, and their embryology, in detail to see whether their similarity really is fundamental, or superficial and analogous. Homology represents a powerful concept for morphological organs such as wings. Wings are complex in structure and can take on an almost infinite variety of shapes; they have an embryonic development and morphological relations with the rest of the body. If the information in the structure and development of a wing in two species is the same, it is highly likely that the character evolved from a common ancestor who had similar wings.

The homology/analogy distinction is much less powerful for molecular evidence. Suppose a nucleotide is identical (as an *A*) in two species. Evolutionary changes take place among a very limited set of alternatives (the four bases *A*, *C*, *G*, and *T*), and it is fairly probable that the same informational state could independently evolve in the two species. The argument is relaxed for proteins, because 20 amino acid states are possible. Nevertheless, it still applies because 20 fixed states make up a small number compared with the variety of morphological forms. Thus, on the molecular level it is not so unlikely that similarity in the states of two species could have evolved independently. Moreover, the morphologist's methods are absurd for molecules. The amino acid found at site 12 of cytochrome *c* is methionine in humans, chimpanzees, and rattlesnakes, but glutamine in all other species—including many mammals and birds—that have been studied. We cannot dissect the rattlesnake's methionine, or trace its embryonic development, to see whether it is only "superficially" methionine and "more fundamentally" glutamine. It is a methionine molecule, and that is that.

Nor can we usually assess the reliability of different pieces of molecular evidence by considering how natural selection could have acted on them. When morphologists examine a similarity between the organs of two species, they also watch for functional convergences—such as the evolution of wings in species that fly. This kind of analysis is impossible if we do not understand the relation between the structure (the wing) and its function (flight). For molecules, we usually lack this understanding. If we knew, for example, that a change from glutamine to methionine at site 12 of cytochrome *c* made functional sense in certain kinds of animals, then the same kind of arguments applied in morphology could be used for the protein. Otherwise, we must treat molecules in the same way that a morphologist would treat an organ of unknown function.

As a result, research with molecular evidence makes little use of the distinction made in morphological research between homology and analogy. Homologous and analogous similarities can be distinguished for molecules, but by the "posterior" method, after the phylogeny has been inferred. With morphological evidence, we saw that it is sometimes possible to infer strongly that a particular shared character is homologous—and the character may then have great weight in phylogenetic inference. For molecular sequences, it cannot be firmly established in advance that an identical amino acid or nucleotide is homologous. We can observe similarity, but the similarity could just as well be homologous as analogous.

Molecular sequences have other distinctive properties. They provide a great deal of evidence. Cytochrome c, for example, has 104 amino acids that can be treated as 104 pieces of phylogenetic evidence. A typical morphological study might be based on perhaps 20 or so characters, and it is exceptional for many more than 50 characters to be used.

In addition, the recognition of independent units of evidence is apparently straightforward with molecular sequences. With morphological evidence, two apparently separate organs may really comprise a single evolutionary unit. At one extreme, non-independence is obvious; no one would think of treating the right leg and the left leg as two pieces of evidence. Less obvious correlations can also arise as a consequence of developmental processes, which makes the recognition of independence tricky. For nucleotides, the mutations down the DNA molecule are effectively independent; each site can evolve independently of each other site. (This is not true in fact, however, when there is no or little recombination between the sites—though how much of a problem this creates is the business of current research.)

Evolution at different amino acid and nucleotide sites is easily comparable. One change at one site is equivalent to one change at another site. How, then, can we say what amount of evolution in a knee bone is equivalent to any given change in a skull bone?

The four properties of DNA and protein sequence data—the impossibility of any deeper analysis of the character, the large amounts of evidence, the recognizability of independent units, and the comparability of evidence—have encouraged the development of statistical techniques to infer phylogenies. The same techniques are, in principle, equally applicable to morphological evidence, although in such cases it is always tempting to try to preempt statistical analysis and resolve the apparent conflicts by ever-deepening character analysis. Morphological data are also less readily divisible into neat character states for statistical analysis.

Molecular Sequences

A number of statistical techniques can produce an unrooted tree from a set of molecular sequences. They broadly divide into "parsimony" and "distance" statistics. In this section, we will concentrate on parsimony.

The first, and most difficult, step is to align the sequences from the various species. With normal length sequences of more than 100 nucleotides, regions will usually have been deleted during evolution in some species and added to others, such that the sequences of the different species do not simply align, with nucleotide number 39 of species 1 corresponding to the nucleotide 39 of the other species. Alignment is a difficult art, and we shall not discuss it further here.

Given a set of aligned sequences, the principle of parsimony with molecular sequences follows the same procedure as used in the simple example of the human, chimpanzee, magnolia, and amoeba presented at the beginning of this chapter. We write out all possible phylogenies and count the minimum number of events implied by each. Figure illustrates a mechanical procedure for some simplified nucleotide data (although it looks at only three of the seven possible unrooted trees for five species). Four of the five sites in the figure are informative (only site 5 is uninformative, being equally compatible with all trees), a higher proportion than found in most real cases.

At Stage 4, the stage at which we have the most parsimonious unrooted tree, the branches in which character changes occurred are known, but their polarities are not. In the figure, the branches with changes have lines drawn through them. In the top left branch, for example, a change occurred between *A* and *T*, but we do not know whether $A \rightarrow T$ or $T \rightarrow A$. The polarities of the changes are needed at the final stage, at which we find the root of the unrooted tree. If the characters were not molecules, it might be possible to find the root using any of the criteria. With molecular evidence, however, only outgroup comparison can be used. We therefore look for the sequence in a closely related species and see to which branch in the unrooted tree it corresponds most closely.

In one case, the outgroup sequence most parsimoniously branches off the bottom right-hand branch of the unrooted tree, and implies the rooted tree as drawn. In the other case, the root of the tree might be located in any of three branches. It is also possible that outgroup sequences might be unknown or so different from the five species that further evidence would be needed before we could locate the root of the tree.

Molecular Evidences

The art of using molecules to infer phylogenetic relations involves picking a molecule that evolves at a rate appropriate to the group of species in question. Different proteins, and stretches of DNA, evolve at different rates, and they can be used like clocks with hands that move at different rates. If you use a rapidly evolving molecule for an ancient group, the molecule will have "turned over" many times during the phylogeny. In such a case, once multiple changes at the same site become common, the phylogenetic information in the sequence similarity is lost. By analogy, a stopwatch with only a seconds hand would be of no use in comparing professors' lecture times. Likewise, slowly evolving molecules are useless for fine phylogenetic resolution because they will not have changed enough.

Ribosomal RNA genes are particularly valuable in phylogenetic reconstruction because they are found in almost all species; they are present in both mitochondrial and nuclear DNA. The mitochondrial genes evolve more rapidly than nuclear DNA, and mitochondrial ribosomal RNA genes are useful for resolving phylogenetic problems in the 10-100 million year range. In contrast, the slowly evolving nuclear ribosomal RNA genes are useful in the hundreds of millions of years range.

Thus, when Milinkovitch et al. wished to resolve the phylogeny of dolphins and whales, which the fossil record suggests to have originated less than 35-40 million years ago, the mitochondrial rRNA genes were an appropriate choice. Much of it is uncontroversial, but one of the

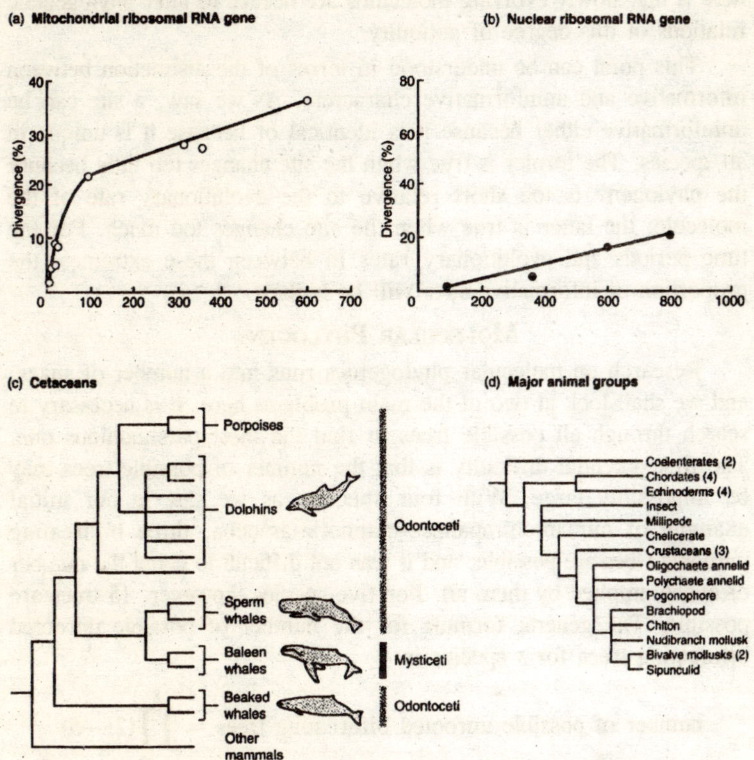

Fig. 7.13. Matching the molecule to the phylogenetic problem.

branches is surprising. Cetaceans are usually classified into two groups, Odontoceti (which contains the echolocating toothed whales and dolphins) and Mysticeti (which contains the filter-feeding baleen whales). Milinkovitch et al.'s mitochondrial DNA evidence suggests that the Odontoceti are paraphyletic and that one group of odontocetes (the carnivorous sperm whales) shares a more recent common ancestor with baleen whales than with the other odontocete groups. The grouping will need further independent confirmation before it becomes generally accepted.

By contrast, the results of a study by Lake on the grand groups in the animal kingdom. These groups originated hundreds of millions of years ago, so the nuclear rRNA genes were the appropriate molecule for the problem. Careful inspection of the phylogeny reveals another zoological surprise—the arthropods may be paraphyletic. Lake's result, at least for the arthropods, has been challenged, but the main point here is that slowly evolving molecules are needed to infer phylogenetic relations of this degree of antiquity.

This point can be understood in terms of the distinction between informative and uninformative characters. As we saw, a site can be uninformative either because it is identical or because it is unique in all species. The former is true when the site changes too little because the phylogeny is too short relative to the evolutionary rate of the molecule; the latter is true when the site changes too much. For the time periods and evolutionary rates in between these extremes, the proportion of informative sites will be highest.

Molecular Phylogeny

Research on molecular phylogenies runs into a number of snags, and we shall look at two of the main problems here. It is necessary to search through all possible trees to find the most parsimonious one. The first potential difficulty is that the number of possible trees may be impossibly large. With four species, as we saw in our initial example of human-chimpanzee-magnolia-amoeba, three bifurcating unrooted trees are possible, and it was not difficult to count the number of events implied by them all. For five species, however, 15 trees are possible. The general formula for the number of possible unrooted bifurcating trees for *s* species is:

$$\text{number of possible unrooted bifurcating trees} = \prod_{i=3}^{s}(2i-5)$$

The Π term means "product": we multiply all possible terms in the parentheses. For three species, $s = 3$ and we have only one term to take the product (from $i = 3$ up to s, which is also 3); the parenthetic term for $i = 3$ is $6 - 5 = 1$, and the number of possible trees is therefore 1. For $s = 4$, we must multiply that 1 by the parenthetic term for $i = 4$, which is 3. Thus, $3 \times 1 = 3$, the number of unrooted trees for 4 species. For $s = 5$, the product is $5 \times 3 \times 1 = 15$, and so on. The number of possible trees increases explosively as the number of species increases. For 50 species, there are about 3×10^{76} possible unrooted trees, and for the 30 million species that may be alive on earth today, the number is about $10^{300,000,000}$. No computer can search through that quantity of trees. Approximately 25 species is the practical upper limit for which we can search all possible trees.

Students of molecular phylogenies distinguish between "algorithms" and "optimality criteria." Parsimony is an example of an optimality criterion. It says that the best tree is the one requiring the least evolutionary change. An optimality criterion is a criterion against which all possible phylogenies can be compared, and the best estimate of the phylogeny is the one that is closest to the criterion. Optimality criteria often encounter the problem of limited computer search capacity, because all trees must be compared with the criterion in this method. If the number of species is too large for all possible trees to be searched, the search must instead be carried out by means of an "algorithm." An algorithm is a rule about how to search from one tree to the next, and to assess which of the two trees is better; it eventually will find a tree that is better than any of the alternatives, but it searches through only a limited number of trees to reach that end.

The following analogy should clarify how a search algorithm works. Suppose you are in San Francisco and are giving someone instructions on how to find Los Angeles. An optimality criterion would be to say "find the city with the largest population in the United States." The unfortunate person who receives this direction must visit every city in the country, and measure their population sizes, to be sure he or she has found that destination. (We assume the searcher has no other source of information.) An algorithm might say "face south and, keeping the Pacific Ocean on your right-hand side, move forward until you arrive at a city with more than one million inhabitants." Now only a small proportion of the United States must be searched, and the conclusion will be satisfactory so long as no other cities exist that meet the criterion between the starting and finishing points.

With the principle of parsimony, the optimality criterion is known, and research is concerned with developing algorithms that search through the trees as efficiently as possible. The particular algorithms used in phylogenetic research have constantly improved in recent years, but remain vulnerable to becoming trapped on "local optima" when they search through the possible trees in a particular way. A local optimum is a tree that appears to be the best possible choice, by comparison with the other trees that the algorithm investigates, but is actually less parsimonious than other trees in a very different part of the space of possible trees. One practical response to the problem is to run the algorithm several times on a set of sequences, starting each run at a different starting point in the "tree space." If all runs converge on the same answer, it strongly suggests that the algorithm has located the most parsimonious tree. If the runs give conflicting results, however, it may suggest that the evidence is inadequate in some way.

A recent study of humans using mitochondrial DNA illustrates this problem. (A mitochondrial type is a particular mitochondrial sequence. Mitochondria were sequenced from 189 individual humans, and because the study had 189 humans and 135 mitochondrial types, each tip of the phylogeny represents one, or a few, individual human beings.) The number of possible phylogenies with 135 tips is astronomic; they cannot all be searched. The result is the output after one run of a parsimony algorithm, and it has a number of interesting properties. For example, it shows that the deepest branch is African; it has African mitochondrial types to one side and a mix of African and non-African mitochondrial types on the other side, implying that the root of the tree was an individual who lived in Africa.

Another interesting result is that the mitochondrial types do not fall into the groupings that might have been expected. Look, for instance, at the Yorubans. The Yoruban mitochondrial types are scattered through the phylogeny, even though they all live in Nigeria. Likewise, Papua New Guineans do not form a discrete group. It is possible that our

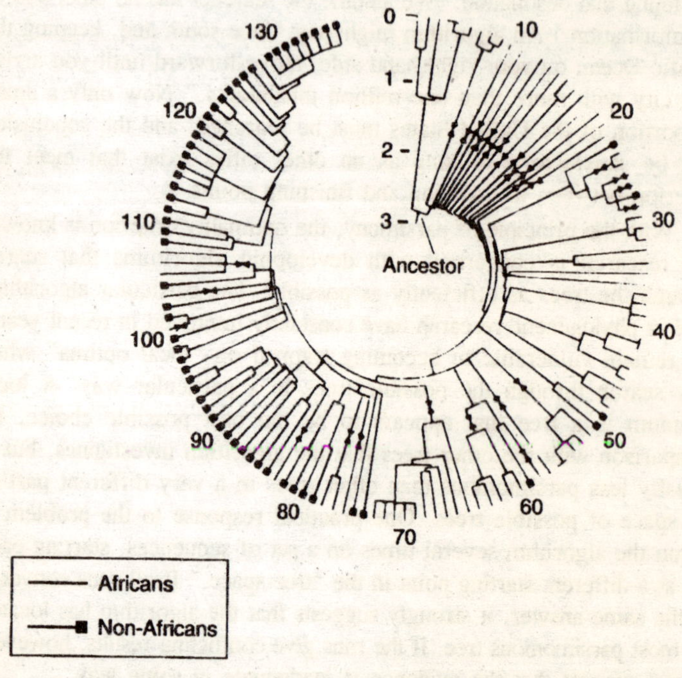

Fig. 7.14. Phylogenetic relations within Homo sapiens, as revealed by mitochondrial DNA.

naive expectations are incorrect—but the results more likely indicate that the tree is unreliable. Indeed, when the program was rerun, starting in different regions of tree space, many more trees were found that were more parsimonious. Some had African deep roots; others did not. In addition, the different trees showed all sorts of groupings of human populations.

The problem of finding the most parsimonious tree was compounded in this study by a further difficulty. It has 135 tips, but the sequence evidence contained only 201 variable sites and only 119 of those sites were informative. Thus, the evidence is insufficient to resolve the tree. As a rule of thumb, at least as many informative changes are needed as there are tips, and ideally many more informative changes should be included. The algorithm might still stop at a local optimum even if the evidence had contained sufficient changes (e.g., 1000 or so) to resolve the tree, but the danger is greater if the evidence is insufficient, because many trees are now equally compatible with the evidence.

The region of the mitochondrial DNA that was used to produce Fig. 7.15 had not accumulated enough changes because it does not evolve rapidly enough relative to the time scale of the problem. The time scale is short, because human populations may have possibly diverged as recently as 100,000 years ago or less. More evidence is needed to resolve the tree, perhaps from a more intensive study of mitochondria, or by turning to other parts of the DNA that evolve even faster. A human phylogeny that Bowcock et al. (1994) constructed using a set of DNA sequences called *microsatellites*. Microsatellites are made up of a series of short, repeating sequences and they evolve rapidly. These high-speed sequences are well tuned to resolve human populations. In summary, we have considered two separate, but related problems. First, when the number of species (or other taxa) at the tips of the phylogeny is large, the number of possible phylogenies may be too large for all of them to be searched. The algorithms that are used to search among the trees are usually reliable, but not infallible. Second, when the evidence is insufficient, the algorithms are particularly likely (on any one run) to become stuck on a local optimum, and fail to find the most parsimonious tree or even the set of all equally parsimonious trees.

Chromosomal Inversions

For some reason, an extraordinarily large number of species of fruitflies (*Drosophila*) are found in the Hawaiian archipelago. There are probably about 3000 drosophilid species in the world, and approximately 800 of them appear to be confined to this archipelago. The phylogeny of one subgroup of the Hawaiian fruitflies is better known than that of any other equivalently large group of living creatures. It was worked out, by Carson and his colleagues, from chromosomal banding patterns. Chromosome bands are clearly visible in fruitflies.

The banding patterns differ between species, and it soon becomes obvious that regions of the chromosomes have been inverted during evolution. In fact, a segment of genes within a chromosome has been inverted as a whole. The important event, for phylogenetic inference, occurs when a second inversion takes place across the end of an earlier inversion. When this modification happens, we can infer with near certainty that the unrooted tree is 1↔ 2 ↔ 3, not 1 ↔ 3 ↔ 2. If species 1 had evolved directly into species 3, and then 3 into 2, the two inversions would be needed for the evolution of species 3. To go to species 2, the exact same two breaks (one at each end) of the second inversion would have to happen again in reverse—which is much less probable than evolution in the order 1 ↔ 2 ↔ 3. As more species are added, with more overlapping inversions, the improbability of most alternative trees multiply to the point of practical impossibility.

To apply the technique, we must first identify the chromosomal banding patterns of the group of species. One species is then picked, more or less arbitrarily, as the "standard" against which the other

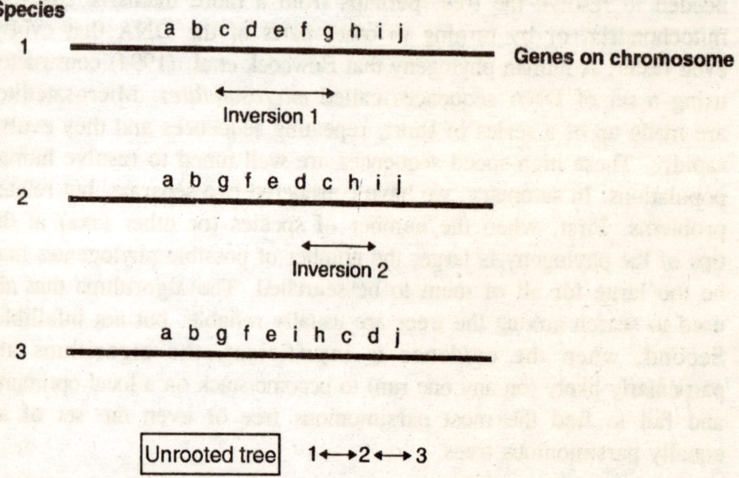

Fig. 7.15. Overlapping inversions, in different species, can be used to infer their phylogenetic relations, in the form of an unrooted tree.

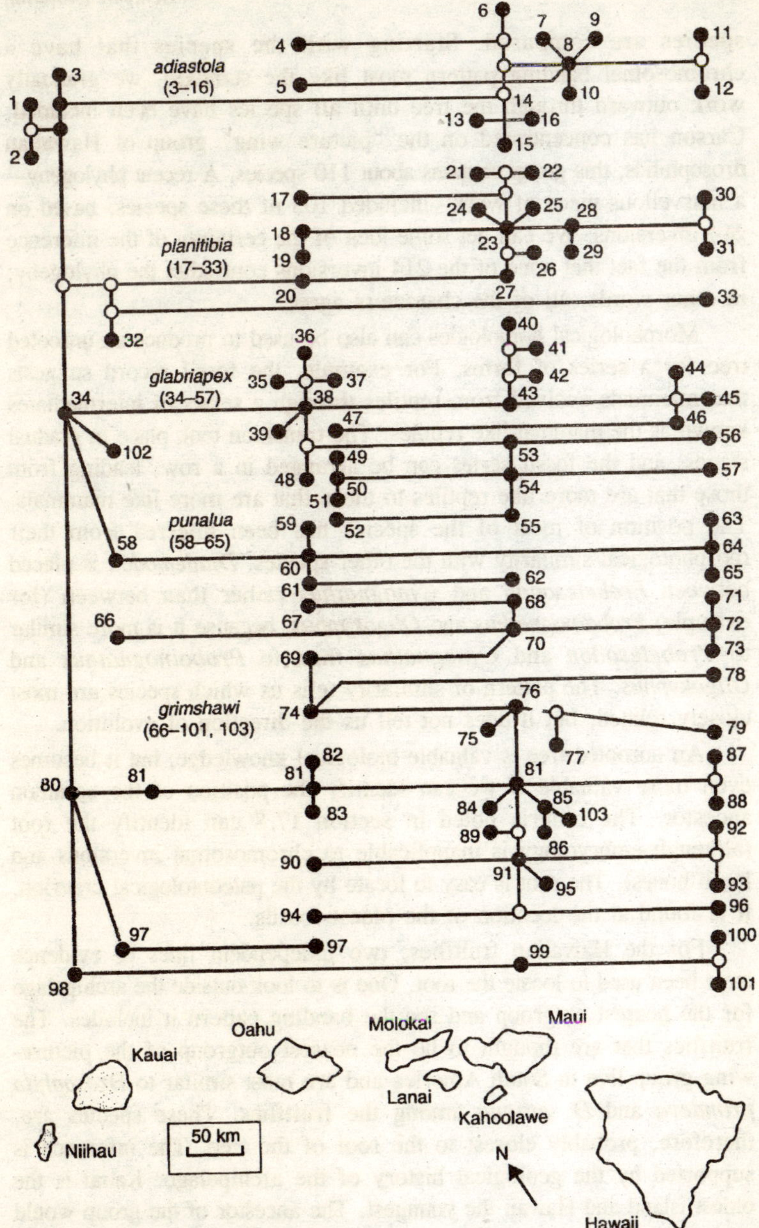

Fig. 7.16. Phylogeny of 103 species of Hawaiian fruitflies (Drosophila) of the picture-wing group.

species are compared. Starting with the species that have a chromosomal banding pattern most like the standard, we gradually work outward through the tree until all species have been included. Carson has concentrated on the "picture wing" group of Hawaiian drosophilids; this group contains about 110 species. A recent phylogeny—a marvellous piece of work—included 103 of these species, based on 214 inversions. We can get some idea of the certainty of the inference from the fact that none of the 214 inversions contradict the phylogeny; in other words, all of the characters agree.

Morphological homologies can also be used to produce an unrooted tree for a series of forms. For example, the fossil record suggests that mammals evolved from reptiles through a series of intermediates known as the mammal-like reptiles. The transition took place in gradual stages, and the fossil series can be arranged in a row, leading from those that are more like reptiles to those that are more like mammals. The position of most of the species has been inferred from their morphological similarity with the other species. *Diademodon* is placed between *Probelesodon* and *Cynognathus* rather than between (for example) *Probainognathus* and *Oligokyphus*, because it is more similar to *Probelesodon* and *Cynognathus* than to *Probainognathus* and *Oligokyphus*. The pattern of similarity tells us which species are most closely related, but it does not tell us the direction of evolution.

An unrooted tree is valuable biological knowledge, but it becomes even more valuable if we can identify the position of the common ancestor. The criteria noted in section 17.8 can identify the root (although embryology is inapplicable to chromosomal inversions and fossil bones). The root is easy to locate by the paleontological criterion. It is found at the location of the oldest fossils.

For the Hawaiian fruitflies, two independent lines of evidence have been used to locate the root. One is to look outside the archipelago for the nearest outgroup and see the banding pattern it includes. The fruitflies that are thought to be the nearest outgroup of the picture-wing group live in South America and are most similar to *Drosophila primaeva* and *D. attigua* among the fruitflies. These species are, therefore, probably closest to the root of the tree. The inference is supported by the geological history of the archipelago. Kauai is the oldest island and Hawaii the youngest. The ancestor of the group would probably have colonized Kauai and, if it is still alive, it should still be on Kauai because almost every Hawaiian drosophilid is confined to a single island. *D. primaeva* and *D. attigua* live on Kauai. Indeed,

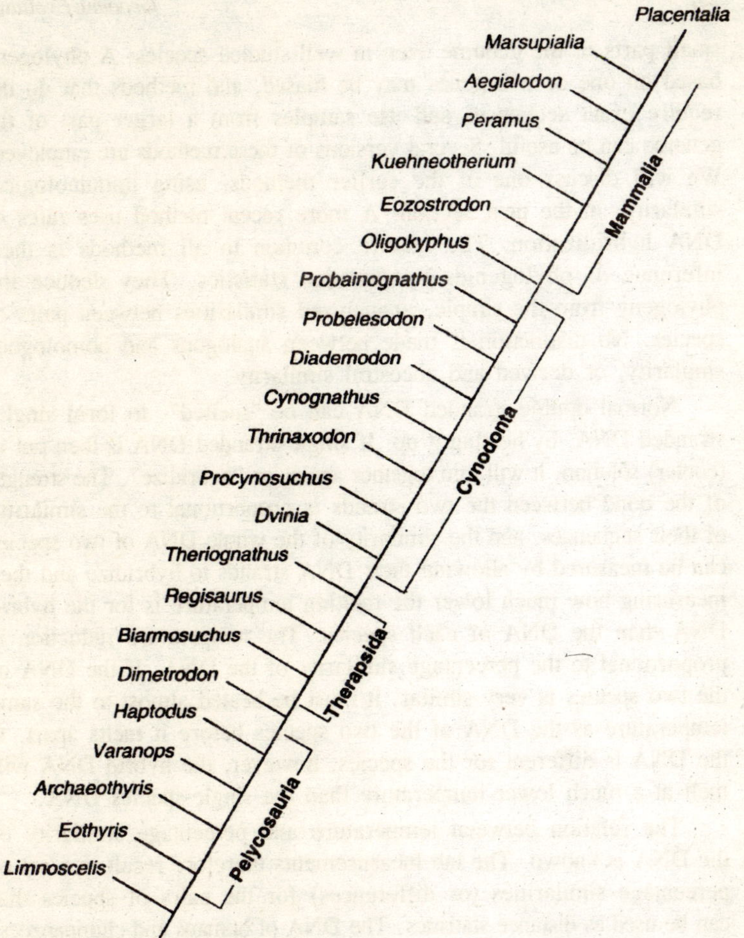

Fig. 7.17. A morphoseries of fossil mammal-like reptiles.

much of the phylogenetic history of the picture-wing group consists of populations of species on the older islands moving to the younger islands, where they form new species by allopatric speciation. No examples are known of species on older islands that are derived from a species on a younger island. Thus, the youngest island, Hawaii, supports the most recently evolved species and the oldest islands in the west have the most ancient species.

PHYLOGENETIC RELATIONS

The rate at which molecules—both proteins and DNA—can be sequenced is increasing constantly, but we still have sequenced only small parts of the genome even in well-studied species. A phylogeny based on one or two genes may be biased, and methods that do not require exact sequences and use samples from a larger part of the genome can be useful. Several versions of these methods are employed. We will discuss one of the earlier methods, using immunological similarity, in the next section. A more recent method uses rates of DNA hybridization. The feature common to all methods is their inference of phylogenies by distance statistics. They deduce the phylogeny from the simple, unanalyzed similarities between pairs of species. No distinction is made between analogous and homologous similarity, or derived and ancestral similarity.

Normal double-stranded DNA can be "melted", to form single-stranded DNA, by heating it up. If single-stranded DNA is then put in (cooler) solution, it will join together again, or "hybridize". The strength of the bond between the two strands is proportional to the similarity of their sequences, and the similarity of the whole DNA of two species can be measured by allowing their DNA strands to hybridize and then measuring how much lower the melting temperature is for the hybrid DNA than the DNA of each species. The temperature reduction is proportional to the percentage similarity of the DNA. If the DNA of the two species is very similar, it must be heated almost to the same temperature as the DNA of the two species before it melts apart. If the DNA is different for the species, however, the hybrid DNA will melt at a much lower temperature than the single-species DNA.

The relation between temperature and percentage similarity of the DNA is known. The lab measurements therefore result in a set of percentage similarities (or differences) for the pairs of species that can be used as distance statistics. The DNA of humans and chimpanzees, for example, differs on average at about 1 1/2 % of nucleotides. This finding is the evidence for the famous claim that the DNA of humans and chimps is 98.5% similar. The figure can be converted into a time to the common ancestor if a species pair can be found for which both the age of their common ancestor and the percentage difference in their DNA is known. The fossil record suggests that Old World monkeys and great apes shared a common ancestor about 30 million years ago, and this figure can be used to convert percentage differences into absolute times.

The problem with distance statistics is that they do not distinguish between the three kinds of shared characters. If shared ancestral similarities are more common between two species than are shared

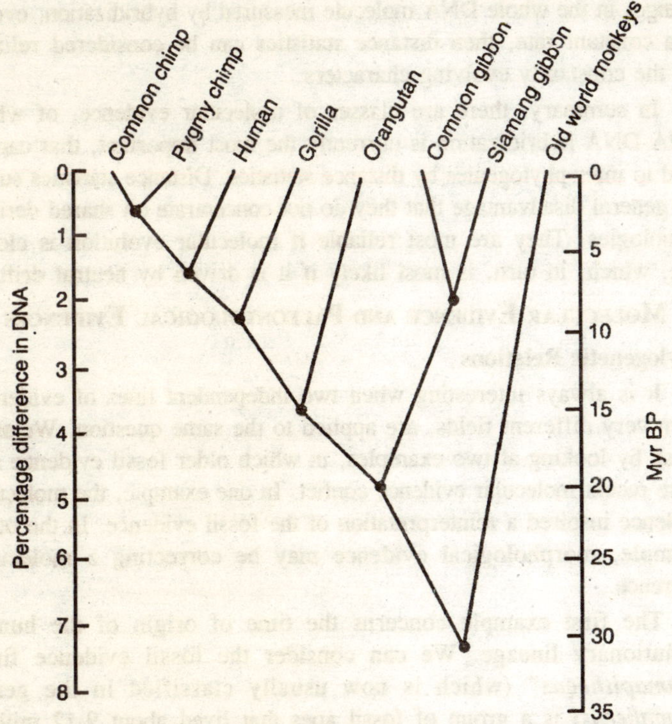

Fig. 7.18. Phylogenetic relations of hominoids, as revealed by DNA hybridization.

derived similarities, distance statistics confuse paraphyletic and monophyletic groups. If some members of a group of species (like the reptiles) evolve slowly and other members of the group evolve rapidly, the slowly evolving species resemble one another even though they do not share a recent common ancestor. Thus, distance statistics group species by similarity and will recognize the disparate set of slowly evolving species as a phylogenetic group.

It can be argued that this general problem does not apply to DNA hybridization. DNA hybridization measures the percentage similarity of the whole (or a large part of) DNA of two species. The DNA sequences of species may well evolve at a relatively constant rate. Genes and proteins provide reasonably good molecular clocks and the same is likely to be true for the DNA as a whole. In that case, the DNA hybridization method is justified. The problematic cases like reptiles arise when evolution proceeds more rapidly in one lineage than another. If that rapid evolution occurs only in some characters, such as observable morphology, while other characters, such as the changes in the whole DNA molecule measured by hybridization, evolve at a constant rate, then distance statistics can be considered reliable for the constantly evolving characters.

In summary, there are classes of molecular evidence, of which DNA-DNA hybridization is currently the most important, that can be used to infer phylogenies by distance statistics. Distance statistics suffer the general disadvantage that they do not concentrate on shared derived homologies. They are most reliable if molecular evolution is clock-like, which, in turn, is most likely if it is driven by neutral drift.

Molecular Evidence and Paleontological Evidences

Phylogenetic Relations

It is always interesting when two independent lines of evidence, from very different fields, are applied to the same question. We shall finish by looking at two examples, in which older fossil evidence and more recent molecular evidence confict. In one example, the molecular evidence inspired a reinterpretation of the fossil evidence. In the other example, morphological evidence may be correcting a molecular inference.

The first example concerns the time of origin of the human evolutionary lineage. We can consider the fossil evidence first. "*Ramapithecus*" (which is now usually classified in the genus *Sivapithecus*) is a group of fossil apes that lived about 9-12 million years ago. Until the late 1960s, almost all paleoanthropologists thought that *Ramapithecus* was a hominid—that is, it was more closely related to *Homo* than to chimpanzees and gorillas. (Hominoids formally superfamily Hominoidea) are the group of all great apes, including humans; hominids (formally family Hominidae or subfamily Homininae) are the narrower group consisting of *Homo* and the australopithecines.) *Ramapithecus* and *Homo* apparently shared a number of derived characters. For example, *Homo* has a rounded, "parabolic" dental arcade, whereas chimpanzees have a more pointed dental arcade. *Ramapithecus's* dental arcade was initially thought to be shaped more like that of *Homo*. *Ramapithecus's* canine teeth were thought to be relatively diminished compared with its other teeth, as in *Homo* but unlike chimpanzees (in which the canines, especially in males, are large). Most importantly, *Homo* and *Ramapithecus* were thought to share, as a derived condition, a thickened layer of tooth enamel, unlike the thinner layer in other apes (and which was thought to be the condition in the ancestors of the Hominidae).

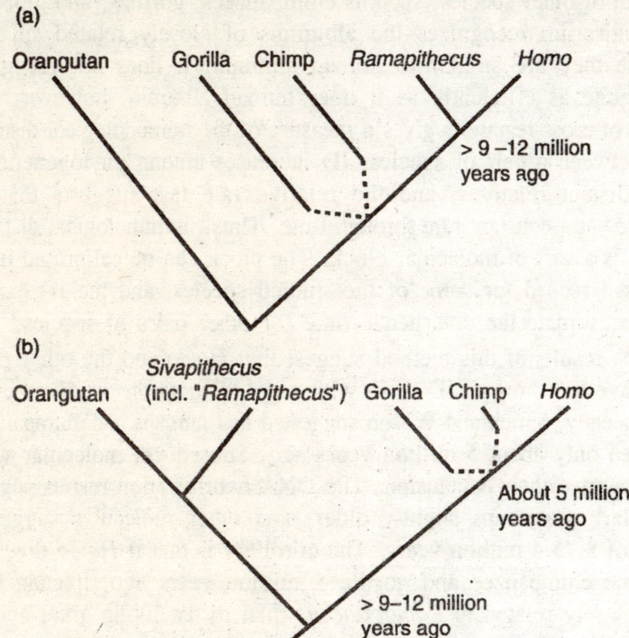

Fig. 7.19. Relations of Homo, other great apes, and Ramapithecus, according to (a) original paleontological and morphological evidence, and (b) molecular evidence.

This morphological and paleontological argument for a relation between *Homo* and *Ramapithecus* has a classical form. A set of character states were shown to be shared uniquely by these two species, and the characters were derived within the larger group of Hominoidea. The corollary was that the human lineage must have split from the great apes at least 12 million years ago, because *Ramapithecus* is nearer to humans than to the great apes.

In the early 1960s, Goodman first demonstrated the molecular similarity of humans and other great apes. The molecular argument for a recent human-ape split was most influentially made, however, in a paper by Sarich and Wilson in 1967. Sarich and Wilson used an immunological distance measure. This method is similar in philosophy to DNA hybridization, but differs in the exact molecule used.

To measure immunological distance, Sarich and Wilson first made an antiserum against human albumin by injecting human albumin into rabbits (albumin is a common protein that circulates in the blood). They then measured how much that antiserum cross-reacted with the albumin of other species, such as chimpanzees, gorillas, and gibbons. The antiserum recognizes the albumins of closely related species, because they are similar to human albumin. It does not recognize them quite as efficiently as it does human albumin, however. The degree of cross-reactivity gives a measure of the immunological distance (ID) between a pair of species. ID increases among phylogenetically more distant relatives, and the relative rate test suggests that ID increases at a constant rate through time. Thus, immunological distance serves as a sort of molecular clock. The clock can be calibrated using the fossil record for some of the studied species, and the ID can be used to estimate the divergence time for other pairs of species.

The results of this method suggest that *Homo* and the other great apes have too short an ID to fit with a pre- *Ramapithecus* divergence. Consequently, Sarich and Wilson suggested that humans and chimpanzees diverged only about 5 million years ago. Subsequent molecular work has supported their conclusion. The DNA hybridization results suggest a similar, if perhaps slightly older, and other molecules suggest a figure of 3.75-4 million years. The corollary is that if *Homo* diverged from the chimpanzee and gorilla 5 million years ago, it cannot be more closely related to *Ramapithecus* than to the living great apes.

Clearly, the molecular and fossil evidence disagreed. A controversy began, in which both the molecular and morphological evidence was challenged (often by experts in the other field). The controversy has now been settled (with a few dissenters) in favour of the original molecular evidence. The morphological characters previously believed to show a relation between *Homo* and *Ramapithecus* succumbed to reanalysis. The dental arcade of *Ramapithecus* had been wrongly reconstructed (originally by combining parts from different specimens), the canine teeth may appear reduced because the fossil *Ramapithecus* specimens were female, and Martin in 1985 finally removed the last important character—thickened enamel—by reinterpreting it as an ancestral character. Moreover, when *Ramapithecus* was compared with another fossil (*Sivapithecus*) that was generally accepted to be a close relative of the orangutan, and with the orangutan itself, it showed clear similarities to them. The specimens formerly classified as *Ramapithecus* are now usually included in the genus *Sivapithecus*, which in turn is thought to be a close relative of the ancestors of modern orangutans.

In summary (simplifying things a little), molecular evidence helped to inspire a reanalysis of the fossil evidence for hominid origins—with the result that a figure of about 5 million years (or at least the range of 4-8 million years) is now widely accepted for the time of origin of the hominid lineage.

Amniote Relations

Some would draw the conclusion from the previous example (and others like it) that molecular evidence is more reliable than morphological and paleontological evidence. The uncertainties in molecular phylogenetic inference that we encountered in this example would certainly cast doubt on that conclusion, however. In this section, we will consider another phylogenetic problem that is the subject of current research and in which the opposite process may be happening. We say "may be" because the controversy is current, and no conclusion has been reached as yet.

This problem concerns the relations of the amniotes (the group of mammals, birds, and reptiles). Mammals and birds are usually thought to have evolved separately, from different reptilian taxa. The main evidence for this suggestion comes from fossils. A group of reptiles called the mammal-like reptiles (or synapsids) is known to form intermediates between the reptiles and the mammals. Another set of fossils, including *Archaeopteryx*, connects the birds with the diapsid reptiles, a very different group.

In 1982, Gardiner suggested a different phylogeny, using the morphology of modern amniotes. He considered 47 morphological characters and concluded that birds and mammals form a group (called the Haemothermia, because they are warm-blooded) that evolved from the reptiles. The Haemothermia would share a common ancestor subsequent to the reptiles, whereas in the pattern of their common ancestor would be found deep within the reptiles. If the fossil evidence and Gardiner's morphological evidence are accepted at face value (which not every one does), the two conflict. If the birds and mammals are closely related groups, the fossil evidence for their separate origins must be reinterpreted to show that it is misleading in some way. On the other hand, if the fossil evidence is accepted, we must show that the modern morphological evidence is misleading, perhaps because the independent evolution of a warm-blooded physiology by birds and mammals led to many other convergences. Gauthier et al. (1988) showed that the traditional grouping is supported when all evidence is included and analyzed by parsimony.

What does the molecular evidence suggest? The answer is not simple, but the weight of evidence at least initially seemed to support

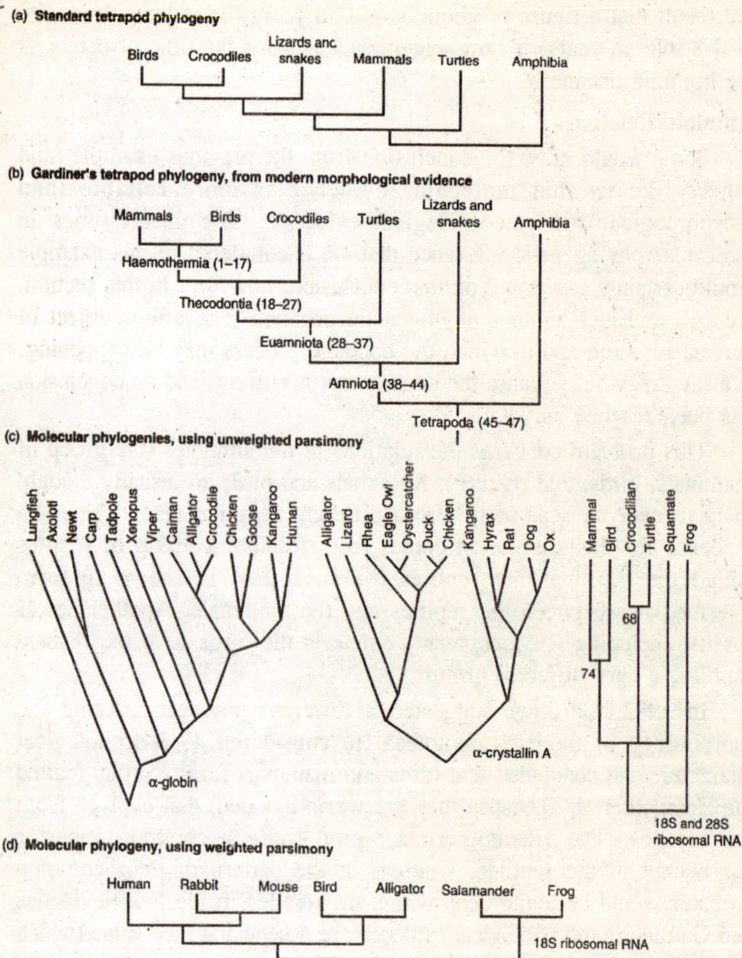

Fig. 7.20. (a) Standard phylogeny of amniotes and their near relatives, based on skull anatomy and fossil evidence from mammal-like reptiles. (b) Gardiner's phylogeny. (c) Molecular evidence.

Gardiner's controversial scheme. α-globin supports α-crystallin A does not. The ribosomal RNA genes support using one method of phylogenetic inference (but not when using another, arguably inferior, method). Perhaps this question represented yet another case in which dusty fossil evidence was to be brushed aside by go-go molecules. In this instance, the comparative anatomy of living species, according to Gardiner's analysis, supported the molecular evidence. Although that assessment may yet turn out to be correct, current research does not appear to be heading that way. The fossil evidence in this case (unlike with *Ramapithecus*) is extensive and cannot easily be ignored or interpreted away. The force of the fossil evidence has driven people to look harder at the molecular evidence.

We saw earlier how parsimony can be used in various "weighted" and "unweighted" forms. The DNA of birds and mammals is known to be G-C rich, however. That is, it has a higher proportion of the nucleotides G and C as opposed to A and T when compared with reptiles and other taxa. The reason behind this variation may be the warm-blooded physiology of birds and mammals, and their similarity in G-C content would then be likely to be analogous rather than homologous. We can test whether the high G-C content of mammals and birds influences the result by analyzing the pattern suggested when we use a form of weighted parsimony in which A-T similarities are weighed more heavily than G-C similarities.

These results will not be the final word in this case. It is not known, for example, whether the particular set of weights is correct, and ribosomal RNA is not the only molecule that can be investigated. We can draw the provisional conclusion that molecular evidence, appropriately weighted, may well support the traditional tetrapod phylogeny. It is worth noting that the weightings added to parsimony effectively reintroduce the morphologist's distinctions into molecular research. We noticed earlier that such distinctions are often ignored for molecules, mainly because we do not know enough about molecules to distinguish reliable from unreliable evidence. As our knowledge improves, the distinctions can be reintroduced.

In summary, the phylogeny for amniotes suggested by fossil evidence was challenged by evidence from modern morphology and molecules in the 1980s. Instead of crumbling in the face of the newer analytical method, it helped to force a reanalysis of the molecular inference. The statistical technique used initially with the molecular evidence was probably misleading, and a new system of weights has been introduced. Although only time will prove which results are correct, it appears as if the evidence of dead bones may be correcting the molecular research on amniote relations.

Conclusion

Phylogenetic inference draws on all kinds of evidence, from molecular sequences, to chromosomal inversions, to morphology in modern and fossil forms. It employs (in some form or other) the parsimony criterion: the best estimate of the true tree is the arrangement requiring the least change. With morphological evolution, the most common (though not universal) course of research is to distinguish analogies from homologies, and try to infer the polarity of the homologous characters. The conflict between the characters shared among the species may then be reduced (ideally to zero) and a residue of reliable shared derived homologies used to infer the (rooted) tree. With molecules, individual characters are analyzed less and parsimony is used to infer the unrooted tree; the location of the root can then be inferred by other evidence. Molecular evidence can also be analyzed, and weighted parsimony used in an attempt to focus on the evidence from the more reliable parts of molecules. In other cases, molecules are thought to evolve in enough of a clock-like manner for distance methods to be used as a short-cut to the phylogeny.

It is easy to be deceived by discussions of phylogenetic inference into thinking that it represents an exceptionally uncertain, shaky kind of science. In a full discussion of an unsolved and controversial problem, there can seem to be an endless series of pieces of evidence pointing first one way and then another. One student, when confronted with a classically recalcitrant problem in phylogenetic inference—the origin of the chordates—summed it up as "Paleontology is mute, comparative anatomy meaningless, and embryology lies." This impression could be misleading, however. Discussion (as well as research) naturally focuses on unsolved problems—and these issues tend to be the difficult ones. Many phylogenetic problems have been solved, and it is worth balancing our reasonable certainty in the human-chimpanzee-magnolia-amoeba case with a more slippery problem such as the amniotes. In addition, the phylogeny of the picture-winged fruitflies of Hawaii, deduced from 214 conflict-free, multiple overlapping, chromosomal inversions, the phylogenetic inference has a level of certainty that compares favourably with most of the facts of the natural sciences.

8

RECENTLY EVOLVED GENES

Discoveries of the elaborate structure of genes raise evolutionary questions that lie outside the scope of traditional theory. Darwinism rationalizes how structures evolve, but does not address the more significant issue of the roles that specialized gene structures play in the process of evolution itself. I suggest that molecular genetics can be integrated into evolutionary theory only through the paradigm that underlies all biological explanation; the structure-function principle. Organisms carry out evolutionary processes as biological functions. To do so, species evolve special structures with *evolutionary functions* just as they evolve other phenotypic structures to carry out various adaptive functions. I shall discuss the evolutionary functions of a variety of structures, ways that these functional capacities evolve, and some implications of evolution being a process that organisms become specialized to carry out instead of just change imposed on the species from the external environment.

Heredity lies at the heart of all theories of evolution. Indeed, evolution is the time dimension of genetics. Darwin's greatest handicap in explaining evolution was a complete ignorance of genetics. This forced him to accept two interwoven causal processes in evolutionary change. Natural selection drove evolution by survival of the fittest but an internal hereditary process generated the individual variability on which selection acts. Darwin was unable to evaluate the relative importance of these two agencies. As time went on, he admitted an increasing role for heredity in directing the process of evolution through some sort of Lamarckian mechanism. The discovery of Mendelian genetics rescued natural selection. Recombination of particulate genes, and ultimately blind mutation, produced continuing variation. Since both processes could be completely random and spontaneous, natural selection no longer had to be diluted with organismal hereditary activities. The stark objectivity of the new Darwinism allowed it to overwhelm its vitalistically tainted competitors. Its causal simplicity also bedazzled the 'Neo-Darwinist' into believing that the simplest possible way that evolution could take place must also be the way that evolution actually did proceed.

This misconception might have been exposed by the growing doubts from the fossil record as to whether Neo-Darwinism addresses the key features of the actual history of life on the planet. In fact, it was insight into heredity that again proved pivotal. The revolution of molecular biology has made it possible to examine directly the structure and physiology of genes. Genetic determinants turn out to be very different from the ineffectual abstractions upon which Neo-Darwinism was based. We still cannot foresee their full role in evolution, but it is probably substantial. I suggest that evolution is primarily an internal genetic process. Natural selection, as we know it, occurs and is essential, but it is the junior partner to heredity. Not only the mechanisms but also the goals of evolution are defined by the genetic message. To describe this interpretation I will first sketch the main characteristics of genes as functional structures, then describe the observed evolutionary dynamics of genes, and finally suggest that the phenotype evolves through these same categories of changes.

Nature of Gene

Genes are Units of Molecular Structure

The modern geneticist extracts genes as molecules from cells, looks at them with the electron microscope, measures their physical properties, such as length, and describes their chemical composition. This is entirely different from the approach that prevailed when Neo-Darwinism was being formulated. T. H. Morgan observed in his 1933 Nobel lecture that: 'There is no consensus of opinion amongst geneticists as to what the genes are—whether they are real or purely fictitious—because at the level at which the genetic experiments lie it does not make the slightest difference whether the gene is a hypothetical unit, of whether the gene is a material particle'. Since then, the main concern of genetics has shifted from inheritance of genes to their structure. This approach is fruitful because genes are surprisingly elaborate in structure and this structure is the basis for their function.

Genes are Highly Individual

Until this decade, the quest of genetics was for universal truths about inheritance—Mendel's laws, Fisher's fundamental law of evolution, Crick's central dogma. This made the simplest genes the ones of choice for study. The resulting lowest common denominator orientation reduced genes to no more than runs of codons in DNA molecules, distinctive only in their sequences. Such forms of genes do exist, and probably are abundant, but they seem to be used only for elementary coding functions. Typically, genes with complex information and active evolutionary dynamics are elaborate, and their significant capacities stem directly from their individualistic structure. This realization has a profound evolutionary implication. To understand how important phenotypic traits evolve we must describe the actual dynamics of the complex genes involved, and not merely the stereotyped behaviour of simple genes.

For example, it once seemed all but impossible that the mammalian genome could code for hundreds of millions of different antibody molecules that encompass every possible foreign antigen. Yet vertebrates are able to do so and even make antibodies against artificial molecules to which their species had never been exposed in its entire evolutionary history. Both the capacity to code for the immune response and the ability to evolve this information are made possible only by the very special and unique structure of the immune genes.

Even lowly bacteria rely on highly distinctive gene structures. Although the run of the mill genes in bacteria have a stable configuration on the chromosome, those genes that are actively evolving tend to develop into configurations such as transposons. These structures can move around in the genome and can even hitch-hike from one cell to another by associating with other specialized DNA structures called transmissible plasmids. By exchanging transmissible genes among other species in the ecosystem instead of relying on mutations, bacteria are able to evolve far more rapidly and effectively than would be possible through simple mutation and selection.

It took decades for bacteriologists finally to accept the empirical data that bacteria evolve resistances to antibiotics far too fast for a process of mutation and selection and it has taken evolutionists even longer to appreciate how micro-organisms actually do evolve. As late as 1970, a treatise on evolution still insisted that 'Microorganisms... have the capacity to develop strains resistant to antibiotics and other drugs. This resistance results from the selection of a few resistant mutations of gene combinations exactly as in higher organisms'. With the growth of modern genetics we now know that this is a secondary mechanism for bacterial adaptation. Adaptive evolution in bacteria does not centre on mutation and selection. It is foremost a process of restructuring genes into transposable and transmissible forms and then moving them about the cell and the ecosystem. Thus, for example, the important information for evaluating the potential of a bacterial pathogen to evolve resistance to an antibiotic is not the mutation rate or the frequency of resistant individuals, but the *structure* of the genes that provide that characteristic.

Genes are Working Molecules of the Cell

The third cornerstone of modern genetics is what I call the 'profane' side of the DNA molecule. Classical genetics was predicated on the genotype having a sacred status. Genes were inviolable messages from ancient ancestors passed on faithfully from generation to generation, except for rare mutational corruptions. They were transcendental to the activities and metabolism of the organism carrying them. We now realize that this simply is not true. Instead cells actively manipulate the structure of their DNA molecules for both physiological and evolutionary reasons. They engineer changes in their genes with an impressive array of 'gene-processing enzymes'. Cells seem to have enzymes to catalyze almost every imaginable sort of change in the structure of the DNA molecule. Topoisomerase unties knots in DNA strands, recA protein inserts single strands of DNA into homologous double-stranded helices, acquisitionase grabs bits of DNA and intercalates them into chromosomes and so on. These enzymes for manipulating DNA molecules are not only diverse but also numerous. For example, there are thought to be at least 20 different sorts of DNA glycosylases.

Complex genes, such as multigene families, owe much of their distinctiveness to special enzymic pathways that manipulate their DNA sequences. For example, the antibody and T-cell receptor multigene families have highly specialized enzymic pathways to rearrange their DNA in particular somatic cells and the duplicated genes in the haemoglobin families are enzymically 'corrected' one against another in germ line cells during evolution.

The ultimate authority over the information content of its DNA would be for the cell actually to write gene segments from scratch, and to edit the result in order to achieve a useful product. Remarkably, even this capacity is realized! Mammals are able to polymerize *de novo* sequences of nucleotides at crucial interior sites in the genes for their antibodies. Those physiologically written segments code for particular portions of antibody molecules that recognize their antigens. The creations that are useful are selected from others by an impressively complex multicellular system. A decade ago, geneticists would have considered this capacity unimportant for evolution because it is confined to a small number of highly specialized genes. It is not a general property of genes. However, we now realize that every important gene is probably exceptional in one way or another, and that it is these unique genetic mechanisms that have allowed the important accomplishments of life to evolve.

The full contribution of gene-processing enzymes to evolution remains uncertain. One obvious role is to produce mutants. Probably the majority of mutations important in evolution result from the activities of enzymes. This is significant because enzymes are deliberate and precise, the antithesis of the supposed 'randomness' of mutation. Indeed, enzymically induced changes in the structure of multigene families often are referred to as gene 'switching', 'correction', 'rectification', 'excision', and so on, rather than 'mutation'. I also find it suggestive that geneticists have chosen names for gene-processing enzymes that could be used almost as well for the keys on a word processor.

Genes are Aware of their Surroundings

Genes require information about the organism and its environment in order to express their message anpropriately. To meet this demand, complex bacterial genes (operons) have evolved special sensory proteins called 'repressors', which inform them of particular relevant conditions. A typical repressor has two binding sites, one specific for some meaningful metabolite in the cell and a second for a recognition site in the operon. The two binding capacities are mutually exclusive, so that the repressor tells the operon about the concentration of the metabolite by binding to the gene's DNA. Sensor proteins of bacteria are versatile. Some activate rather than repress genes, and some inform genes about hormonal signals generated by the bacterium, or about the status of other genes in the genome. Also, some operons receive information from several sensory systems simultaneously.

Eukaryotic genes are more complex and less well understood than those of bacteria. They are tucked away in a membrane-bound nucleus, and germ line cells are isolated in gonads from the soma. Despite these specializations, receptor proteins still carry hormone molecules directly to sites on the chromosome within the nucleus. Even germinal cells of vertebrates produce receptors on their cell surfaces for a variety of hormones, growth factors, neuromodulators and so forth. Higher organisms presumably have a far greater potential than bacteria for informing their genes but we still know almost nothing of its evolutionary significance.

Types of Evolutionary Change in Genes

In addition to revising our ideas about the fundamental nature of the gene, molecular geneticists also have discovered the types of change that occur in genes during their evolution. Each of these has been chosen as a definition for 'evolution' by one school or another.

Fluctuations in Gene Frequencies

Fifty years ago it was known that genes were located on chromosomes and could mutate to alternative states. Their structure was otherwise mysterious. The only information about genes that could be interpreted (other than map positions) concerned the mathematical *frequencies* of their detectable allelic forms in populations. The obvious basis for a genetic theory of evolution was change in this measurable quantity. Neo-Darwinists therefore defined evolution as simply a change in frequencies of gene alleles in a population from one generation to the next. Such quantitative change is fully reversible for a gene, and the frequencies of individual alleles can shift back and forth to adapt organisms in the population to their changing environment. New alleles occasionally arise by mutation or become 'fixed' by completely replacing their alternative alleles but these processes are very slow and irrelevant to immediate adaptation. In fact, species evolve numerous and diverse mechanisms especially to minimize both the loss of variation by gene fixation and the formation of new mutations, most of which are deleterious. The Neo-Darwinian process concerns fluctuations in the frequencies, but not the forms, of genes.

Change in Gene Structure

When it became technically possible to read the amino acid sequences of primary gene products (and later the nucleotide sequences of the genes themselves), a new paradigm for molecular evolution became the comparison of homologous sequences taken from related species. Species were found to diverge from one another, and from their ancestors, by the accumulation of small substitutions in the molecular structure of their genes. Irreversible permanent change in DNA structure replaced population dynamics as the definition of gene evolution.

'Evolutionary distance' is now measured as the number of substitutions accumulated in the structure of a gene or protein with time. Curiously, this yardstick gives anomalous readings of relatedness among species. It measures the length of time since two species diverged from a common ancestor rather than how much their biology has changed during that time. This is because most base substitutions that accumulate in genes have little effect on phenotype. They are selectively neutral, or nearly so, and driven inexorably into the gene by mutation pressure. Comparative biochemists recognize the distinctiveness of this predominating process of relentless gene change by calling it 'non-Darwinian evolution'. This implies that Darwinian adaptation through natural selection is a separate, and quantitatively lesser, process of change. The main debate in molecular evolution today is about the relative contributions of neutral and adaptive processes to change in gene structure.

Adaptation

There have been two main approaches to the study of adaptive changes in gene structure. One is to expose laboratory populations of micro-organisms to artificial selection pressures and then to analyze the changes by which they adapt. The other is to identify the functionally important parts of genes or proteins and to describe how these vary among related species.

In pioneering the former method, John Langridge spread large numbers of *Escherichia coli* on agar that contained lactobionic acid as the only energy source. This sugar is a derivative of lactose, a disaccharide that *E. coli* metabolizes using the enzyme β-galactosidase. However, the enzyme cannot hydrolyse the analogue sugar effectively. Mutant colonies were readily obtained and genetic analysis showed that many of them had genetic alterations in their β-galactosidase gene.

Langridge's major discovery was the striking variety of *alternative* ways in which the genes were able to acquire this new function. The adapted state is not a unique one, and when two populations or species adapt to a common environment they can be expected to do so in different ways. This raises a key unresolved question. What determines how a gene will evolve when multiple alternative pathways for adaptation are available? The factors that decide the choice between alternatives probably are as important in determining what will evolve as are the environmental stresses to which the species is adapting. So far, few inquiries about gene evolution have reached the degree of sophistication necessary to address this question.

The other approach, of examining the particular parts of macromolecules responsible for functional properties, takes advantage of molecular biology's strong emphasis on structure-function relationships in macromolecules. It is now possible to picture protein molecules in three dimensions and to identify the local regions responsible for substrate binding, catalysis, subunit interaction and so on. Comparing these particular parts of a protein between species shows that adaptation of genes is distinct from just change in structure. Perhaps the most significant revelation from these studies is that the more functionally important a segment of gene is, the slower it changes. In fact, the principal way to locate the functional elements in a gene or protein is to search for regions that have not changed during evolution. This means that selection acts mainly *against change*—a defensive role quite distinct from the Neo-Darwinian view of selection as the driving force *for change* in gene frequencies. Natural selection has been likened to an editor whose main activity is to proofread for errors rather than to promote change.

Advancement

Genes do slowly acquire adaptive changes, but it is clear that not all such changes are of equal importance. Although the majority of selected mutations optimize structural details for the exact functional demands of the environment, a smaller number *advance* the organization of the gene. They create functional capacity instead of adjusting it, and the gene or protein is developed, not just adapted.

Haemoglobin illuminates this process. Most adaptive alterations in this protein probably adjust its oxygen-binding curve to compensate for changes in the organism's size, habitat and so on. In addition, a smaller number of entirely different sorts of changes have *advanced* the haemoglobin molecule from a primitive monomer to its derived tetrameric form. This oxygen-binding protein originally consisted of a single polypeptide chain with a haeme group, a form that still persists in myoglobin. When the chains evolved the capacity to aggregate, they acquired the possibility for allosteric interactions between subunits. This allowed oxygen-loading curves to become sigmoidal with a threshold and a narrowed range between the oxygen tensions for loading and unloading. Allostery also permitted oxygen loading to be regulated by concentrations of other ligands, especially hydrogen ion and diphosphoglycerate. Allosteric regulation is essential to the function of modern haemoglobin, and to enzymes in general. Indeed, nearly all enzymes with sophisticated functions have evolved into oligomers. In the evolutionary history of haemoglobin, the emergence of the capacity for allostery was a landmark advance distinct from its subsequent adaptive fine-tuning.

Several other developments further advanced haemoglobin. One was the evolution of two different types of subunits, alpha and beta, which join together to form the molecule. Heterodimerism allows subunits individually to evolve specialized functions. The importance of this capacity again is evident from the large number of other proteins which also have dissimilar subunits.

Another profound advancement was the organization of the haemoglobin genes into multigene families. Mammals have tandem arrays of genes for the alpha-type and beta-type subunits. The gene copies in these families are differentially expressed during ontogeny to match the haemoglobin molecule to the individual demands of the embryo, foetus and adult. Some species have developed additional haemoglobin gene copies for various other special purposes. Clearly, the vertebrates have actively utilized the opportunities for adaptation opened up by the organization of haemoglobin genes into multigene families. However, the significant evolutionary step was that which created this special adaptability in the first place.

The distinction between advancement and adaptation stands out in many genes, especially in those with demanding biological functions. Significantly, the two processes depend on different kinds of mutations. Genes adapt mainly through base substitutions and occasional small deletions. In contrast, they advance by a wide variety of rearrangements and additions as well as by point mutations. For example, the genes for antibodies acquired their advanced structure by capturing transposons, shuffling exon gene segments, developing special enzymes for rearranging and mutating gene sequences and repeatedly duplicating segments of genes, whole genes and entire multigene families. Molecular geneticists interested in structure-function relationships repeatedly have found it important to distinguish the process that originates and develops structure from the lesser process of its subsequent adaptation. Adaptation obviously plays a role in the origin of novelty, but the two processes are categorically different.

Principles of Evolutionary Advancement

We have a reasonable understanding of how genes fluctuate in allele frequency, change in structure, and adapt. If progression of form is distinct from these processes it is important to determine just how it occurs. Four extra principles seem to be emerging from studies of gene structure. They are discussed here in order of increasing causal complexity.

Tautology of Advancement

Structures that are the most advanced are those that were able to advance the most. This really says no more for evolution than it does for a handful of peas spilled on the ground. Some peas will roll further than others and there must be a reason for the variation. Three can be imagined: (1) some peas might be intrinsically better able to roll, perhaps due to a rounder shape or smoother skin; (2) the surface of the ground may vary from place to place, blocking some peas with bumps and rough spots and facilitating others with depressions or grooves that channel their velocity; (3) chance variation as peas hit the ground and jostle one another will move some peas further than others. The analogous factors in evolution are (1) the genetically determined organization of the species, (2) the selective environment, and (3) whim.

Obviously it is important to evaluate which of these factors have caused some forms of life to advance so much further than others during their 3×10^9 or so years of existence. However, the interactions between phenotype and ecology of higher organisms seem to be vastly too complex for resolving this issue. Fortunately, advancement at the level of gene structure is more tractable. Here, an important observation is that genes that have evolved conspicuously in a particular direction often have identifiable physiological mechanisms that enhance the ability to change in that direction. This association is strong enough for molecular geneticists to rely on it to discover significant physiological mechanisms. For example, a number of multigene families with outstandingly large numbers of gene copies are associated in some way with special mechanisms for reproducing gene copies. These include 5S RNA genes, rRNA genes, pseudogenes, IAP sequences of mice and other retrovirus-like genes, transposons, VAT genes of trypanosomes, possibly satellite DNAs, and unique sequences intercalated into satellite DNA.

As other examples, the extraordinary uniformity of gene copies in some multigene families imply special enzymic mechanisms for removing mutational variation, and one can predict that specialized genes that occur in similar form among unrelated species of bacteria probably are organized as transposons or associated with transmissible plasmids to aid their dissemination. A species also may progress notably in a particular evolutionary direction because it has overcome some factor that constrains such evolution in other species. This represents another way by which a gene or species can specialize its phenotype for a particular kind of evolution. There is a good deal of speculation about structural constraints on the evolution of macromolecules but neither clear examples nor a comprehensive analysis of this issue has yet appeared.

Preadaptation

A particularly important problem for evolutionary advancement concerns the origin of new form. Once a structure advances enough to be functional it can develop further by selection; but how does it initially develop enough to begin to function? This problem worried Darwin and has yet to be resolved. Two principles are probably important. One relates to the well accepted concept of preadaptation. A structure that had evolved originally for one function subsequently acquires a second function. Selection for the original function elaborates the structure until its side effects happened to become useful in other ways. The swim bladder, from which the lung arose, and the reptilian foreleg, which evolved into the avian wing, are familiar preadaptations. The term preadaptation today represents only the opportunism of the evolutionary process without any of the teleological overtones implied by certain earlier evolutionists.

Genes are particularly favourable as preadaptations because they can be duplicated by a single mutation. One copy can preserve the original function, while the other is unhindered to acquire a new function. Moreover, a supernumerary gene can become silent and drift to a new structure unimpeded by selective constraints. Gene duplication and divergence have been dominating processes for evolving new proteins. Most proteins of eukaryotes are members of families that arose from single ancestral genes. These genes of common descent are homologous in the same sense that the wing of the bird is homologous to the leg of the reptile. Of course, not all duplicated genes acquire a new function. The genome of higher organisms is rife with redundant gene copies that have failed to acquire a new function and are degenerating back to random sequence as 'pseudogenes'.

It is significant, but not surprising, that when a duplicated gene acquires a new function that new usage is usually related to the original one. Some groups of homologous enzymes span a wide range of catalytic activities, but most related genes have related functions. This suggests that when genes for a function proliferate, the preadaptive potential for further advances of that general function may increase as well. Some genes have spawned many more new functional derivatives and pseudogenes than others, but it is not clear whether this reflects structural features of the genes or just the functions that they serve.

Autopoiesis

A second mechanism for the origin of novelty involves the physical principle of autopoiesis. This is the spontaneous emergence of organization out of nothing—or, more precisely, out of noise. Closed physical systems and open ones near equilibrium gravitate towards definite simple states of maximal entropy and minimal rate of entropy formation, respectively. Dissipative open systems far from equilibrium have entirely different behaviour. They can spontaneously became more organized with time. A system that is highly dissipative gives fleeting existence to an astronomical number of random or chaotic local microstructures. Any of these twinkling organizational states has the opportunity to stabilize itself from decay and to extend its organization catalytically past its borders if it is able. Just a microscopic seed of such 'autopoietic' organization in an energy dissipating system may grow to macroscopic dimensions, typically as a periodic structure. For example, a thin layer of liquid traversed by a large heat flux characteristically develops a regular periodic pattern of convection that is stable whenever the temperature gradient is applied. Analogous self-forming organization also develops in appropriate dissipative chemical systems and is conjectured to emerge spontaneously even in economic and social systems.

Autopoiesis seems particularly applicable to evolution, which generates new self-propagating organization in a highly dissipative system. A major difference between biological and physical processes is that biological systems are extremely organized to begin with in contrast to the typical system for a physics experiment, such as a container of liquid, on a heat source. The physicist's demonstrations are important to show that dissipating energy does have the capacity to organize matter spontaneously but the appropriate examples to clarify the role of autopoiesis in evolution must come from biology and not thermodynamic paradigms.

Autopoiesis plays an essential generative role in various forms of genetic organization. For example, multigene families easily evolve once they get started because after a gene duplicates it has a propensity to vary further in copy number due to non-reciprocal crossing over. Moreover, the larger a family grows, the more intrinsically variable it becomes. If every additional copy in a multigene family were as difficult to achieve as the first or second, one never would expect to see huge multigene families. However, immense families are common and evolve extraordinarily rapidly. These satellite DNAs give every indication of being products of a runaway autopoietic process.

Autopoiesis has an even more obvious role in the evolution of transposons. A large number of genes scattered along chromosomes and bathed in nucleoplasm rich in enzymes represents the substantial level of organization that characterizes a genome. For example, most cells code for transposases that can synthesize extra copies of a gene and plant them elsewhere in the genome. Brewer's yeast uses a transposase to change sex by copying a silent gene for one or the other sex into a special location where it is expressed. The two sex-determining genes can be copied because they are flanked by special short DNA sequences that the transposase enzyme recognizes as its substrate. Other organisms use transposases for other specialized functions. As long as a transposase, or other gene-processing enzyme, is unrelated to the genes that it acts upon, its activities are causally simple. However, it is possible for the continual shuffling of genes along the DNA molecule to sandwich a transposase gene between recognition sequences for its own enzyme. The enzyme now acts *self-referently* on its own genetic determinant and, in the case of a transposase, a transposon is born. The transposon propagates autonomously by copying itself. Moreover, since its components are genetically specified, the unit can mutate and evolve. Variant copies accumulated in the cell or gene pool can compete with each other for resources, differential fecundity, or *lebensraum* with the 'fitter' form supplanting the others. This evolution is quite distinct from that of other genes because it is *selfish*. Self-reference can allow a gene to evolve for its own sake without dependence on the environment of the organism. Transposons evolve both by modifying the genes they already have and by acquiring new genes.

'Selfish evolution' has been likened to a Darwinian process, with the transposon enjoying the status of an individual 'organism'. Actually the process is considerably more powerful because the 'environment' for the gene is not the inanimate physical world but a milieu determined by a genetic programme. We generally believe that organisms adapt to their environment. However, since transposons can (and actively do) accumulate relevant genes of the host, they are able to adapt their environment to suit themselves. For example, if the amount or specificity of an enzyme in the cell is suboptimal for transposition, a transposon need not adapt itself to that restrictive condition. It can pick up the gene for that characteristic of the 'environment' and selfishly evolve it. Once the gene becomes part of a selfish element it will automatically evolve to optimize its contribution to the fitness of that element.

The transposons that have been characterized (mainly from bacteria) include elaborate forms, attesting to the creative power of their autopoietic evolution. Accumulating such genes does not merely complicate the transposons. It also gives them new emergent properties. Most notably, transposable elements in bacteria have an inherent tendency to become increasingly autonomous of the host. It is believed that genes shuffle actively from one level of autonomy to another as an important process in evolution of bacteria and possibly of eukaryotic genes as well. Thus autopoiesis can create new levels of functional activity and can transfer the determinants of evolution from the external environment to the internal evolving structure.

Evolutionary Function

Some changes in genes are so deliberate that they can only be understood through the biological concept of function. We are well acquainted with adaptive functions, those deliberate biological activities that are carried out to aid survival. In addition, various enzymes that operate on genes have *evolutionary functions*. They do not enhance the fitness or survival of the organism, but instead they facilitate its evolution.

An example of an evolutionary function is the rectification of gene copies in a multigene family. Some families contain dozens or hundreds of gene copies with the same function. It is impossible for the individual copies to evolve directly by natural selection because the redundancy makes the contribution of any one copy totally dispensable. For example, only the most dominantly deleterious mutation in a 5S RNA gene of *Xenopus* could possibly be sensitive to selection with 10,000 sister gene copies in the family. In order to expose sequences in a multigene family to natural selection, organisms must change large groups of the genes to a common structure. If they rectify a block with variant genes to wild type, mutations are purged. If they expand a mutant configuration, natural selection can then favour or eliminate that variant allele of the family within the species.

Some multigene families maintain extraordinary levels of homogeneity. Hundreds or thousands of gene copies are kept almost identical even though they readily evolve in concert. Obviously such uniformity can result only from a very effective rectification mechanism, of which eukaryotes have several. A gene copy can be directly corrected to the sequence of another in some multigene families. Saltatory expansion and contraction, sister chromatid exchange, and extrachromosomal replication and integration probably occur in others. These mechanisms are prerequisites for using multigene families as genetic determinants. Without them not only would a multigene family be unable to evolve its functions, but it would inexorably erode into individual genes. The important point is that rectifying a multigene family promotes its evolution but does not increase fitness. (In fact, the process generates genetic load.) The function of the process is an evolutionary one and not adaptive.

The transposases of bacterial transposons are another set of enzymes with important evolutionary functions. Moving transposons around in a cell does not enhance the cell's fitness at all. In fact, it is probably deleterious on the whole because occasionally a transposon gets inserted into an essential gene. However, transposition is essential to bacterial evolution.

Enzymes are not the only structures with evolutionary functions. Sensory machinery also can regulate cellular processes for evolutionary ends. Repressor systems direct most transposase genes to be active only when appropriate for evolution. Some bacteria and higher organisms have evolved very elaborate sensory systems to induce generalized heritable alterations in their genomes in response to stresses.

Since structures with evolutionary functions have entirely different roles in evolution from those with adaptive functions it is useful to have a specific name for them. Elsewhere I have suggested the term 'evolutionary driver' as the counterpart of the noun 'adaptation'. An evolutionary driver is a genetically determined structure with a function of propelling or steering the evolution of a gene, phenotypic trait or species. Enzyme pathways that rectify multigene families, transposases and their repressor systems, the recA system for inducible mutagenesis in bacteria and so forth, are examples of evolutionary directors which participate in the evolution of genes. If trends continue, we will probably discover a diversity of other evolutionary functions during the next few years. Most of the special organizational features of eukaryotic genes that have surprised geneticists during the past several decades are suspected of having evolutionary functions.

Although the concept of evolutionary function extends some of our ideas about evolution, it retains the conventional notion of 'function'. If the function of the enzyme maltase is to hydrolyse maltose, then surely the function of transposase must be to transpose segments of

DNA, and that of acquisitionase is to intercalate bits of foreign bits of DNA into the chromosome. We would call an effect of a structure its 'function' if it provided part or all of the selective advantage for the development or maintenance of that structure. Organisms can evolve structures with evolutionary functions through the same mechanism of *selection by consequence* that underlies the evolution of adaptive structures. All selection is based on the consequences of a structure rather than the structure itself. Species with greater capacity to adapt to their changing environment will benefit from the consequences of that capacity and can eventually prevail.

As discussed above, preadaptation and autopoïesis greatly increase the ability to evolve new adaptive functional capacities. Most evolutionary drivers probably evolved from adaptive precursors that acted as preadaptations. The repressor on transposon Tn917 is a case in point. Its adaptive function is to repress the erythromycin resistance gene of the transposon when erythromycin is absent from the environment. The repressor also has acquired, probably secondarily, control over the expression of the transposase gene. As a result, the presence of erythromycin induces the erythromycin-resistance transposon to move in the cell in addition to expressing a drug-resistant phenotype. Also, many enzymes used for DNA repair have additional functions in recombination and mutagenesis.

Perhaps the foremost discovery about evolution to come from modern genetics is that organisms carry out evolutionary activities as biological functions. Just how many genes have evolutionary functions is unclear, but the number may be large. Eukaryotes may actually have a greater amount of DNA concerned with evolution than with phenotypic fitness! A wide variety of structures with evolutionary functions should come as no surprise considering the wealth of adaptations that even the simplest species have evolved. Any adapted structure must be considered a possible potential preadaptation for an evolutionary driver. Thus the more a species or gene advances, the more opportunities it has for recruiting adaptive structures into evolutionary roles.

Evolution of Phenotype

As described above, genes change in four distinct ways with time. The frequencies of variant forms fluctuate reversibly in populations, structural alterations become fixed, genes acquire adaptive modifications and they advance in their structure. It is significant that each of these dynamics has its counterpart in the evolution of phenotype.

For example, the distribution of body sizes of European moles fluctuates from one year to another depending on the weather, although there is no evidence for a lasting change. Harsh winters selectively eliminate larger individuals, but milder years allow them to reaccumulate in the population. Presumably body size, like most quantitative traits in animals, is maintained as a long-term balance between intermittent pressures for increase and for decrease. The overwhelming preponderance of phenotypic changes rapid enough to be seen from generation to generation almost certainly are of this nature and leave no mark on evolution over geological time. They are just 'noise' in the long-term process. Most selection on gene frequencies is probably of the stabilizing variety that leads to long-term constancy instead of change in phenotype.

Permanent change in phenotype also does occur, of course. It can be inferred from comparative biology and in some cases can be followed directly in the fossil record. The difficulty is not in detecting change, but in distinguishing changes with no selective importance from adaptations. Obviously changes may be adaptive even though we may not be able to figure out their usefulness. Conversely, and more disconcerting, the selectionist's exercise of devising possible adaptive explanation for a change does not prove that that or any other selective factor caused its evolution. A variety of mechanisms including allometry, mutation pressure, founder effects, fabricational noise, pleiotrophy, drift of various sorts, and gene introgression can cause a phenotypic character to change neutrally or even deleteriously. In recent years, important evolutionists have cautioned against the mistake of assuming that any existing character or phenotypic change automatically must be adaptive. Also, some palaeontologists believe that most morphological change occurs as part of the process of speciation. The role of selection in the formation of new species is poorly understood.

Probably the best guide for picking out the adaptations among evolutionary changes is common sense. The change in the colour of the pepper moth (*Biston betularia*) in England to a darker hue where industrial soot has blackened the tree trunks on which they hide is reasonably inferred to be adaptive. The red colour of blood probably is not. The colour patterns of mollusc shells are more problematical and may include both selective and non-selective factors.

It is easier to document selection for changes over short periods of time than long ones. However, the gulf between the 'ecological time' of the Neo-Darwinist and 'geological time' is so great that there is little reason to believe that the proportion of changes that are adaptive will be the same for the two. It is not even clear which time scale of change is the more constrained by adaptation. One reason for the current interest in the evolution of gene and protein structure is chat it should be so much easier to distinguish adaptive from neutral change where relationships between structure and function are simpler. Yet, even here, selectionists and neutralists bitterly disagree.

Progressive advancement to higher forms is also apparent in phenotypes of organisms. A dominant pattern in the long-term history of life is the appearance of successively higher and higher forms. This does not mean that all lineages inexorably advance. They decidedly do not. The pattern is for most species to get trapped in stasis and for only a minority to continue to advance. Nevertheless, by almost any criterion for 'advancement' that one might choose, life has repeatedly progressed from one record height to another during evolutionary history with the highest forms existing today. Such advancement of phenotype probably involves all four of the principles discussed above for genes.

Tautology of Advancement

The truism that the most advanced forms of life come from the lineages that were able to advance the fastest and farthest suggests that higher organisms evolve by different, and more effective, sorts of changes than do more primitive forms. This is indeed the case. Higher forms evolve in ways that were difficult or impossible for earlier forms, A striking progression has been for life to adapt through modifications first in enzyme catalysis, then in morphology, and ultimately in behaviour. Primitive organisms still evolve largely by changing their biochemical pathways. The main adaptive strategy of bacteria, for example, is to meet changes in their environment with new enzymic activities. To this end they have developed the complex system of transposons and plasmids discussed above. Eukaryotes, in contrast, evolve most notably in their morphology, generating new relationship between cells instead of new types of enzymes. Higher animals continue to evolve the occasional new enzyme pathway, for example, a mammalian lactose synthetase for making milk sugar, but a striking discovery from comparative biochemistry is that *Escherchia coli* has essentially the entire repertoire of basic enzyme activities of human cells. By and large, vertebrates, mammals, man and insects are reflections of special developmental pathways. Morphology offers vastly greater scope for change than do enzymes and, from what we know, is easier and quicker to evolve.

Behaviour allows even more varied and rapid adaptation. The most advanced animals evolve largely through their behaviour. For example, ants are counted among the higher arthropods. Their morphology is not conspicuously advanced but their behaviour makes them outstanding. Ants build large architectural structures, conduct agriculture with both plants and animals, cooperate in societies based on altruism, and take slaves. These functional capabilities emerged from advances in behaviour, although with support from specialized morphology and biochemistry.

Another progression is in the source of variation on which selection can operate. Early forms of life depended on spontaneous mutation. Their advanced descendents developed special enzymic pathways for editing alterations in the DNA. Bacteria evolved the more effective capacity to steal new genes from neighbouring species. Eukaryotes advanced in a different direction, evolving a diploid sexual cycle that generates continual quantitative variation by recombination.

Thus, in various ways, the forms of life that have advanced the farthest in phenotype are commensurately endowed with the greatest capacities for evolving.

Preadaptation

This concept is so familiar in classical evolution that it requires little comment here. Most new functional capacities of higher organisms probably originated opportunistically from pre-existing phenotypic structures.

Autopoiesis

Autopoiesis fits less comfortably with current evolutionary thought because it ascribes the cause of change to endogenous structure instead of interferences from the external world. This is a mode of explanation that Neo-Darwinists have explicitly rejected (although it has gained some recent advocacy from followers of so-called non-equilibrium evolutionary theory). Neo-Darwinism arose in large measure as a repudiation of competing interpretations that evolution was directed from within, such as aristogenesis, orthogenesis, evolution by law, mutationism and so forth. Nevertheless, autopoiesis and self-reference are integral both to the general process of evolution and to specific cases. In fact, the very origin of life was an autopoietic event, with its chicken-and-egg conundrum that faithful replication of the nucleic acid molecule is a prerequisite for the evolution of genetic information, but information for polymerase activity could evolve only after nucleic acids could be faithfully replicated.

Sexual selection is a more specialized type of autopoietic evolution. If female peacocks tend to choose males with large tails as mates, then the genes promoting this attractive masculine character will enjoy a selective advantage. As a consequence, genes that enhance the attraction of females to large-tailed males also will be at an advantage. They will prompt females to mate with the favoured males and hence produce male offspring that are more desirable. Circuits of genes that favour themselves through sexual, social or adaptive processes can erupt into runaway evolution unrelated to economy and ecology. Nevertheless, the directions of such evolution obviously are not random. They must depend in complex ways upon which organizational features of the species are potentially most sensitive to snowballing autopoiesis, and upon the availability of triggers to set the self-reinforcing cycles off. It is suggestive that certain aspects of morphology, such as wing bars, tails and head crests, seem especially prone to exaggeration as sexual adornments of some birds and it is interesting to consider the cause of this pattern.

Far more important self-reference emerges from the action of the organism on its own environment. Neo-Darwinism depends on a Newtonian-like distinction between species on the one hand, and its surroundings as an independent agency on the other. Yet, as Lewontin (1982, 1983) has stressed, this is not realistic. The characteristics of an environment are not independent of the species that inhabit it, but are intimately shaped by the occupants. Probably the most important selective feature of the environment for any higher species is the species itself. Thus, selection pressures are not just dictates of the external world but emerge in large part as the manifestations of the genetic programme of the species itself. This means that the genotype plays two roles in evolution. It responds with adaptive change and it also generates the cause of that change. What a genotype adapts to is largely its own expression. This self-referent cause and effect increasingly dominates evolution as organisms become more competent to manipulate their environments.

Since even the most doctrinaire evolutionist cannot ignore this self-reference completely, Neo-Darwinism has developed special branches of 'frequency-dependent' and 'density-dependent' selection. These have given rise to some really exciting insights into evolution, such as the 'Red Queen paradox' and 'evolutionary stable strategies'. Yet, the most conspicuous feature of deterministic mathematical sojourns into self-reference is their overwhelming complexity. Predictability, optimum and determinism become subordinate to instabilities, alternatives, critical points, feedback and oscillations. If there is one overriding lesson to be learned from the evolution of transposable genes it is that the process of evolution has access to dynamic relationships far beyond the trivia of Newtonian cause and effect. The tautology of advancement implies that significant evolutionary progress is a flag for the more powerful forms of evolutionary cause. I suggest that the important issue is not how the phenotype changes due to the effect of evolution, but how it has become specialized to cause evolution.

Evolutionary Function

An organism's most powerful mode of action is to carry out important activities as biological functions. Recent discoveries about the physiology of genes prove that evolution is able to produce structures to enhance the evolutionary process as well as to aid immediate survival. Much of the phenotype of higher organisms probably functions as evolutionary drivers. For example, the half of eukaryotic organization concerned with sexuality, recombination and mutation did not evolve to enhance the survivability or fecundity of the individual, but for its evolutionary function.

The evolution of sexuality poses a problem for the strict adaptationist because it is a burden on the fitness of the individual. We know that it can be dispensed with because occasional species ranging from plants to insects and vertebrates have become asexual. Scrapping sexuality for asexual reproduction probably aided short-term fitness. Nevertheless, higher organisms have preserved the burden of sexuality except for these sporadic cases because asexuality probably leads to *evolutionary* blind alleys.

Recognizing that phenotypic structures can have evolutionary functions is a separate issue from identifying the ones that actually are evolutionary drivers. I have already discussed the difficulty of sorting out adaptations from selectively neutral changes. Compounding this difficulty is the third possibility of structures evolving to perform evolutionary functions. Even at the level of genotype it is difficult to distinguish among these three possible reasons for a structure's existence. This difficulty is reflected in the fact that geneticists have proposed all three sorts of explanations for each of the organizational features of eukaryotic genes. The problem of assigning functions to structures becomes further complicated when adaptive structures secondarily acquire roles in evolution. They can then serve both sorts of functions simultaneously. An example cited above of a protein with such double function is the repressor of transposon Tn917 which simultaneously regulates both the erythromycin-resistance gene and the transposase gene of the transposon.

A critical task before us now is to assess the extent to which phenotypic traits of higher organisms have acquired evolutionary functions. For example, how widely have canalization systems for reducing noise in development been recruited to guide evolution? A second essential question is how many features of evolution that we take for granted as 'just the way evolution is' actually are results of selection for that mode of evolution? An immediate case is punctuated equilibrium. Is this a specially evolved mode of evolution? Have species developed the capacity or propensity occasionally to reorganize their genomes on a major scale in order to break out of the stasis that otherwise 'straightjackets' established species? Do species evolve special populational, morphological, behavioural or genetic organization to throw off new species? If so, can we explain how and why those features evolved? Also, what about stasis: is the pronounced tendency of many established species to persist with virtually no change an attribute that evolved from selection?

Relationship of Advancement to Adaptation and Change

Both gene structure and phenotype show the four dynamics listed in order of their dependence on each other. Each type depends on change at some underlying level, but may occur with or without producing any higher level dynamics. For example, adaptation presupposes change either in the structure of genes or in the frequencies of gene alleles in the gene pool. However, neither a fluctuation in the composition of the gene pool nor a base substitution in a gene implies that adaptation has taken place.

Advancement is also predicated upon some sort of change occurring but clearly does not inevitably accompany it. Superficially, advancement might seem to require adaptive change but this seems not to be so. For example, coupling a multigene family to an enzyme system that actively corrects one gene copy against another could be a critical advance in the organization of the family even though it had no adaptive importance. Probably complex interactions between the ecology and the structure of organisms determine how fast and in which ways species advance as they change. It is possible that the environment is more important for phenotypic advancement than for advancement of gene structure, but we have yet to understand what situations lead the species to advance instead of simply shifting 'sideways' from one niche to another.

Rates of Evolution

Descriptions of the molecular evolution of genes, and their extrapolation to phenotype, have two implications for rates of evolutionary change. Most notably, *biological structures* play active roles in evolution and must be major determinants of how fast or slow genes and species evolve. For example, their elaborate structures for rearranging and transmitting genes allow bacteria to evolve antibiotic resistance much faster than otherwise would be possible. Genetiç and phenotypic organization probably also contribute to slowing down evolution in other cases, although we currently know much less about the mechanisms for stasis. The significance of these genetically determined factors is that they are evolved traits instead of just uncontrolled dictates of the outside environment. To the extent that biological mechanisms participate in evolutionary change they allow its rate to be programmed internally as an attribute of the species.

Secondly, molecular studies show that genes, and presumably phenotype, change in at least four entirely different ways during evolution. These types of change are substantially disconnected from one another and can proceed at independent rates. Analyses of evolutionary rates must begin with a clear description of the sort of change being considered and then proceed to the mechanisms that produce that form of change.

Endogenous versus Exogenous Cause for Evolution

The discovery that genes are inherently dynamic instead of passive structures undermines the basic Darwinian proposition that evolution is change forced on the species by the outside environment. The null hypothesis for Neo-Darwinism is that if a species were boxed into an environment so favourable that it exerted no selection pressure, evolutionary change would cease. This is unrealistic. No matter how much a species was environmentally pampered, its genes would evolve. Species are powerfully influenced by their environment, but environmental forces are superimposed on endogenous engines for change and advancement.

The relative importance of endogenous organization and external environment is perhaps the central question in evolution today. It does not have a simple quantitative answer. One reason is that the species influences its surroundings as discussed above. A more important reason is that the evolutionary role of the genetic message varies from one species to another. In particular, genetic constitution becomes increasingly important as the species advances to a higher form.

These inferred dates pose an interesting paradox. If we compare the sequences of the human α_1- and α_2-globins, the split apparently happened very recently. In fact, it may post-date the split between humans and the great apes and it certainly post-dates the split between the great apes and the rest of the primates. In this case, the primates outside the great apes should have only one α-globin. The prediction, however, is false: all the primates have two α-globin molecules. The full taxonomic distribution of the α-globins suggests that the gene duplicated at least before the origin of the mammals (about 85 million years ago), and probably before the split between mammals and birds (about 300 million years ago). Yet the similarity of the α_1- and α_2-globins in humans suggests the genes duplicated about 1 million years ago. Which of the figures is correct? This paradox, which is a common characteristic of evolution in gene families, is called *concerted evolution*.

Gene Family

If each tandem repeat of the ribosomal RNA gene evolved independently by mutation, drift, and selection, the different genes might gradually diverge from one another over time. Different mutations arise in different genes occasionally in the lineages of two species and become fixed. Because the evolutionary changes accumulate independently both in the different genes and in the different species, we might expect that the similarity between the genes in a gene family within a species would be approximately the same as the similarity between copies of the same gene in different species. In the modern species A and B at the top of Fig 9.4, if you examine one gene in the gene family, such as the gene at the extreme left, reveals one difference between it in the two species; there is also one difference between this gene and the fifth and seventh gene within species A.

This structure provides only a rough prediction of how evolution could proceed. The different genes in a gene family within a species have been separate since the duplication, whereas copies of the same gene in different species have been separate since the species split apart. If the duplication occurs in both species, then the speciation event will be more recent than the duplication. Other things being equal, the copies of the same gene in different species should then be more similar than the duplicates within a species. On the other hand, correlated selection pressures may operate within a species, tending to make the duplicated copies within a species more similar.

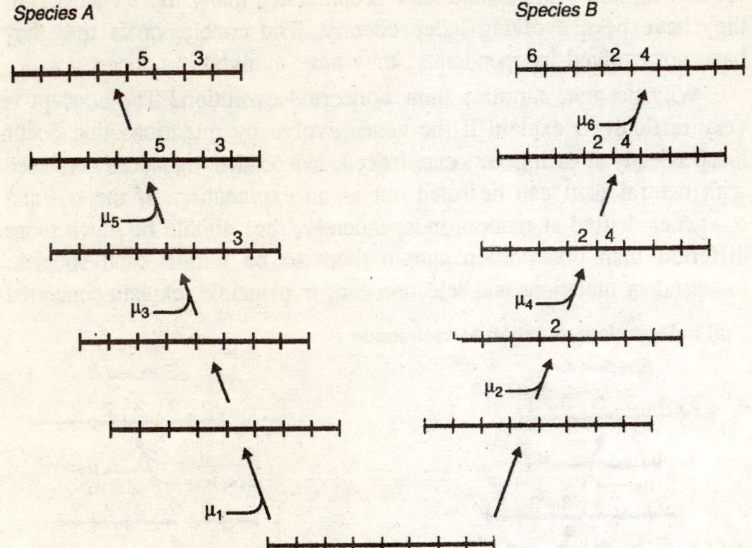

Fig. 9.4. Evolution of a gene family in two evolutionary lineages (in separate species).

And what about the facts? The first study, by Brown et al (1972), was made for the genes that code for ribosomal RNA in *Xenopus*. The 18S and 28S regions both fit the expected pattern. The sequences were almost identical in all tandem repeats of both *X. borealis* and *X. laevis*, suggesting strong stabilizing selection. For the non-transcribed spacer region (NTS), however, the pattern was different. Again, within a species the sequence was similar between tandem repeats, but this sequence differed in the two species. The NTS regions of *X. borealis* and *X. laevis* showed no more similarity than would two quite different genes—whereas within a species the NTS regions all had the same sequence. Thus, the NTS region showed concerted evolution.

The α-globin genes of primates described earlier can be used to illustrate the same principle. All primates, we saw, have two α-globins; we can therefore assume that the common ancestor of primates had two α-globin genes. The sequence of each α-globin gene differs between primate species. In the great apes, for example, any two species differ by about 2.5 amino acid substitutions in each gene. However, within a species, the α_1- and α_2-globins differ by only about one-tenth of that amount (this difference has been shown to be true not only for humans, but also for chimpanzees, gorillas, and rhesus monkeys). If one gene accumulates about 2.5 amino acid changes in the time between two species, then two different genes (α_1 and α_2) that have been separated for 300 million years should have accumulated many more changes—if they have been evolving independently. The conclusion is that they have not evolved independently; they have evolved in concert.

Why do gene families show concerted evolution? The concept is very difficult to explain if the genes evolve by mutations that occur independently at each gene locus. Indeed, independent mutations combined with neutral drift can be ruled out as an explanation. If the α_1- and α_2- genes drifted at random independently, they should be much more different than observation shown them to be within each species. Independent mutations and selection can, in principle, explain concerted

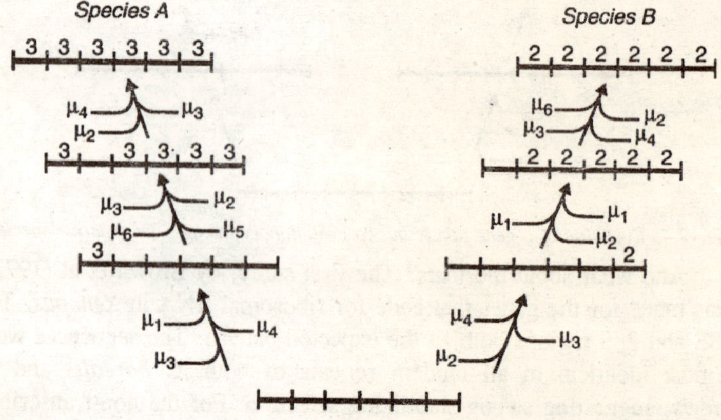

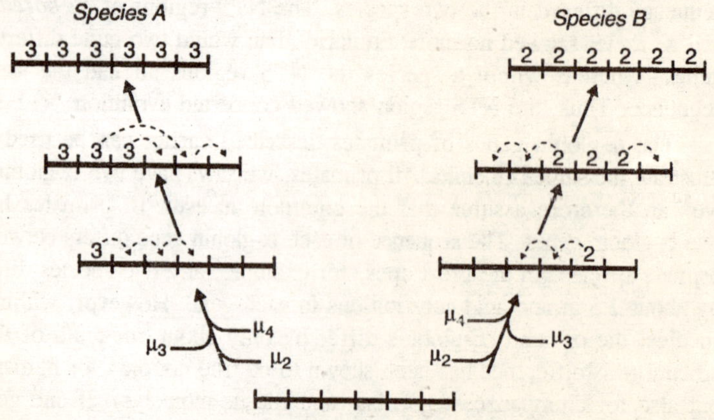

Fig. 9.5. Concerted evolution.

evolution. The same mutations would have to arise independently in both α_1- and α_2-globins in each species and then be fixed by selection, perhaps because one variant of the two genes was favoured in the conditions under which humans evolved, whereas another variant of both was favoured in the conditions under which chimpanzees evolved.

For the two globins in a number of primate species, this theory is a possibility. However, as the gene family showing concerted evolution grows larger, the explanation becomes increasingly incredible. When we reach the ribosomal RNA genes, with their innumerable tandem repeats, it is beyond belief that the same mutation could have occurred independently at each locus and been fixed by selection. In this case, it is generally accepted that some genetic mechanism must act to homogenize the different members of the gene family. One sequence somehow is copied horizontally from one gene to the others. Mutations are not occurring independently at the different loci.

Two main mechanisms have been suggested as causing concerted evolution: unequal crossing over and gene conversion. We encountered unequal crossing over when we discussed gene duplication. The same process generates chromosomes with different numbers and combinations of genes; these chromosomes can provide the mutational raw material for selection or drift to homogenize the gene family. Selection could directly favour more homogeneous gene clusters, or it could simply eliminate clusters with too many or too few copies of a type of gene.

In *gene conversion*, one of the alleles at a locus is converted into the other allele. Thus, when the heterozygote f_1f_2 segregated, instead of the Mendelian proportions of one f_1 for one f_2, two f_1 or two f_2 (and none of the other allele) would emerge. The same process could take place horizontally between the members of a gene family, with the variant at one locus being copied into the other locus. The mutational raw material for concerted evolution would again be produced.

Gene conversion could be either biased or unbiased. Biased gene conversion means that it favours one variant of the gene rather than another. For example, two f_1 genes might be produced more often than two f_2. Biased gene conversion is a kind of directed mutation, and it increases the chance that a gene cluster will be homogenized. Unbiased gene conversion means that the production of two f_1 or two f_2 genes is equally likely.

Unbiased gene conversion and unequal crossing over do not by themselves produced concerted evolution, any more than undirected mutation alone produces other types of evolution. They account only

part of the DNA probably codes for most of the genes. In between is a class of middle-repetitive DNA, which is repeated but not as much as the highly repetitive DNA. On the assumption that the single-copy DNA serves as the coding part of the genome, we can use the proportion of single-copy DNA (i.e., slowly re-associating DNA) as a first estimate of the coding DNA in the genome.

Table 9.1. list the genome sizes and proportion of single-copy DNA for a number of species. These numbers suggest, in two ways, that a large quantity of non-coding DNA exist.

First, the figures for single-copy DNA are often much less than 100%. Only one-fifth of the DNA of the toad *Bufo bufo* is single-copy DNA, for example. If only single-copy DNA acts as the coding part, then four-fifths of *B. bufo*'s DNA is non-coding. This observation is generally true in eukaryotes, but not in prokaryotes. Bacterial DNA appears to be much more economically organized.

Second, large differences appear between species, and these differences can hardly all be due to variations in the numbers of coding genes. Fifteen times as many genes may be needed to build a human as to build a protozoan. In the words of Orgel and Crick, however, "it seems implausible that the number of radically different genes needed in a salamander is 20 times that in a man." The obvious deduction is that the differences derive mainly from non-coding DNA. Thus, a genome consists of more than simply genes.

The experiments provide a second clue about the nature of the non-coding DNA—namely, that it is often repetitive. Geneticists distinguish a number of kinds of repetitive DNA. A first distinction is between *tandem* and *scattered repeats*—that is, whether the repeat sequences appear next to one another (in tandem) or are scattered through the genome. Repetitive DNA can also be classified by the length and number of the repeat units.

No generally agreed classification exists for repetitive DNA, although the suggested classifications are all similar. We will use the following classification developed by Charlesworth et al. (1994). There may be other kinds of repetitive DNA that are not readily classifiable in one of these categories—or there may not be: the question is a topic of current research. (The word "locus" features in the descriptions below. It is used here in a slightly different way from traditional genetics: a locus for a tandem repeat is the site in the chromosome where a whole series of repeats is present. Thus, it refers to the whole array, not just one unit within it.)

Table 9.1. Amount of DNA percentage of single copy DNA in various animal species.

Invertebrates			*Vertebrates*		
Species	*1C (pg)*	*% sc DNA*	*Species*	*1C (pg)*	*% sc DNA*
Protozoa			Protochordata		
Tetrahymena pyriformis	0.2	90	*Ciona intestinalis*	0.2	70
Coelenterata			Pisces		
Aurelia aurita	0,7	70	*Scyliorhinus stellatus*	6.1	39
Nemertini (Rhyncocoela)			*Leuascus cephalus*	5.5	44
Cerebratulus	1.4	60	*Raja montagui*	3.4	47
Mollusca			*Rutilus rutilus*	4.8	54
Aplysia california	1.8	55	Amphibia		
Crassostrea virginica	0.7	60	*Necturus masculosus*	83.0	12
Spisula solidissima	1.2	75	*Bufo bufo*	7.0	20
Loligo loligo	2.8	75	*Triturus cristatus*	21.0	47
Arthropoda			*Xenopus laevis*	3.1	75
Prosimulium multidentatum	0.18	56	Reptilia		
Drosophila melanogaster	0.18	60	*Natrix natrix*	2.5	47
Limulus polyphemus	2.8	70	*Terrapene carolina*	4.1	54
Musca domestica	0.9	90	*Caiman crocodylus*	2.6	66
Chironomous tentans	0.21	95	*Python reticulatus*	1.7	71
Echinodermata			Aves		
Strongylocentrotus purpuratus	0.9	75	*Gallus domesticus*	1.2	80
			Mammalia		
			Homo sapiens	3.5	64
			Mus musculus	3.5	70

Tandem Repeats

1. *Microsatellites*. Repeats of short (2-5 base pairs) nucleotide sequences. The number of repeats varies between loci, but the average is on the order of approximately 100 repeats. Many microsatellite loci are scattered through the genome; an average human, for example, contains about 30,000 microsatellite loci. Microsatellites have also been detected in other vertebrates, insects, and plants.
2. *Minisatellites*. Repeats of longer (approximately 15 base pairs) nucleotide sequences. The number of repeats varies minisatellite loci, and many loci are scattered through the genome. The average length of any one locus is usually about 500-2000 nucleotides. Minisatellites have been studied in humans and other vertebrates, fungi, and plants.
3. *Satellite DNA*. The size of the repeated unit varies in different cases, with some being as small (5-15 base pairs) as micro- and minisatellites, while others are larger (about 100 base pairs). They are often found in large blocks of 1000 or more repeats of the unit sequence in regions of the chromosome near the centromere or the telomere. They do not seem to be as variable in the number of repeats per site as are micro- and minisatellites.

Scattered Repeats

Longer sequences (on the order of approximately 100 base pairs) that are distributed throughout the genome, usually in single copies bounded by other sequences rather than in tandem repeats. Three examples from humans are the sequences called *Alu*, *Kpn*, and poly (C-A); the three together make up about 20% of the human genome.

A unit sequence of any of the three kinds of tandem repetitive DNA can be found at more than one place in the genome. At all those sites they are found in the form of tandem repeats. Scattered repeats are also found at many sites, but usually with one a single copy of the sequence appearing at each site.

Two main evolutionary hypotheses have been suggested to explain the presence of this repetitive, non-coding DNA. One hypothesis says that such DNA is functional, even though it does not actually encode genes. It may be needed for some regulatory or structural reason, for example, or perhaps it keeps the genes apart or correctly configured in the DNA molecule's three-dimensioned shape. Alternatively, the repetitive DNA may be selfish DNA—either neutral "junk" DNA, or parasitic DNA.

We will concentrate on the "selfish DNA" hypothesis here, for two reasons. First, it is widely accepted as the explanation for much of the repetitive non-coding DNA. Second, it is a new evolutionary concept. If the repetitive DNA is functional, it will evolve just like genic DNA and requires no special discussion. If the DNA is non-functional for the organism that contains it, something different is occurring. It is also worth noting that comparisons of non-coding repetitive DNA made between different species have often shown the presence of concerted evolution, as in the gene families of coding DNA discussed earlier. The mechanisms presumably operate in much the same way in the two cases, and we will not discuss concerted evolution here.

SELFISH DNA

In 1980, Doolittle and Sapienza, and Orgel and Crick, made more explicit an idea that many other geneticists had also pondered—the idea that some, or all, repetitive DNA may have no use for the organism. It may instead be *selfish DNA*. Selfish DNA is non-transcribed and non-coding, and contributes nothing to the well-being of the organism. In most cases, such DNA is selectively neutral except for the energetic burden of replicating it. If it was excised, the organism would suffer no disadvantage. After a stretch of selfish DNA originates, it is passively replicated and passed on from parent to offspring. Changes in its frequency in the population would occur by drift. If a non-coding sequence was not neutral—perhaps because it interfered with the construction or metabolism of the organism—it would be selected against. Likewise, if it accumulated to such an extent that the cell cycle was slowed by the need to replicate the entire DNA sequence, selection would probably act to reduce it. Provided the quantity of a particular sequence is not excessive, is not transcribed, and accumulates in parts of the genome where is does not interfere with genic regulation and transcription, there is no reason why selfish DNA should not evolve.

Selfish DNA might be either "active" or "passive". The sequence itself might not influence the chance that it spreads in the DNA and is retained. It is then a passive kind of selfish DNA and could accumulate as "junk DNA" in the genome. Alternatively, a particular sequence might, for some reason related to its sequence, have a better than average chance of spreading through the DNA. These sequences would be a more active, parasitic kind of selfish DNA, and they would proliferate until checked by natural selection because they single copy and the other in multiple repeats—then (other things being equal) the repeated sequence would be more likely to grow laterally through the genome during evolution. If the highly repetitive rows of tandem repeats are actually selfish DNA, their lengths will fluctuate through evolutionary time. Unequal crossing over, slippage, and drift will constantly be causing them to shrink and grow. At any one time, some very long regions of repeats will be found, and these regions will be recognized as highly repetitive DNA. After a few million years, a currently long repeat could potentially have shrunk and some other, previously shorter series lengthened. The frequency distribution of lengths will likely remain probabilistically constant.

Minisatellites (Short Repeats)

While studying the human myoglobin gene in the early 1980s, Jeffreys discovered a short sequence of repeated DNA within an intron. He used the short sequence as a "probe" to see whether the same sequence was present anywhere else in the genome. The probe hybridized at several regions. Jeffreys then extracted the DNA from all these regions, and found that each DNA sample consisted of a unit sequence in a row of tandem repeats. The number of repeats at each site was highly variable between individuals; one site, for example, might contain 25 repeats in one individual, 5 in another individual, and 47 in another. Although the unit sequence itself is variable, a common core sequence has been characterized and is 16 (or perhaps 8) bases long. The particular sequence is not the only sequence that can form this sort of scattered short set of repeats, but it is the most studied. Jeffreys called these sequences minisatellites; the name (like

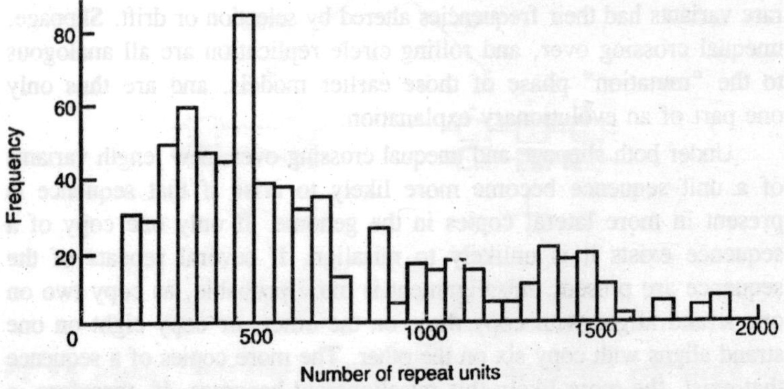

Fig. 9.7. Frequency distribution of numbers of repeats at one minisatellite locus in a total of 344 individuals (688 gametes).

that of microsatellites and satellites) derives from the way they can be detected when DNA is centrifuged.

Why do minisatellites vary so much in the number of repeats? We saw earlier that slippage and unequal crossing over are thought to generate new length variants, and we can now look at two predictions of this idea. Both concern the heterozygosities of different minisatellite loci. The alleles at a minisatellite locus are the different length variants, which differ in their number of repeats of the unit sequence. We can measure the frequencies of the different length variants at each locus, and calculate the heterozygosity for each by the standard formula.

The first relation to consider exists between the heterozygosity of a locus and the similarity of the unit sequence. So far we have described repetitive DNA as if the unit sequence were identical in all repeats. In practice the sequence can differ slightly in terms repeats, with greater differences appearing at some loci. When Stephan and Cho plotted the heterozygosity against the sequence similarity for 10 minisatellite loci in humans, they found a strong negative relation. The loci that are most variable in numbers of length variants possess the most similar unit sequences. The relation is expected if new variants arise by unequal crossing over, for two reasons. First, unequal crossing over is most likely if the unit sequences are more similar. Second, unequal crossing over causes the same variant to be spread through the DNA at a site. If unequal crossing over has happened more frequently at one locus than at another, the unit sequences should look more similar at those loci where the process is more common. If unequal crossing over has not occurred much at a locus, the sequences of the repeats will have diverged more.

Fig. 9.8b shows another relation with heterozygosity for different human minisatellite loci. The more heterozygous loci have higher mutation rates. The mutations in this case form new length variants. If a locus is fixed for one length variant, unequal exchange is unlikely to operate. The more length variants present at a locus, the more likely it is that two different ones will cross over unequally and produce a new variant. That is, the more variety introduced into the system, the more variety it is likely to generate. If the mutations arise by a more conventional genetic mechanism, there would be no reason to predict this trend; if they arise by unequal crossing over, we would expect this result. The mutation rates were estimated by tracing minisatellites through human pedigrees, to see when new variants arose. Notice that the rate is high, up to 5% per gamete.

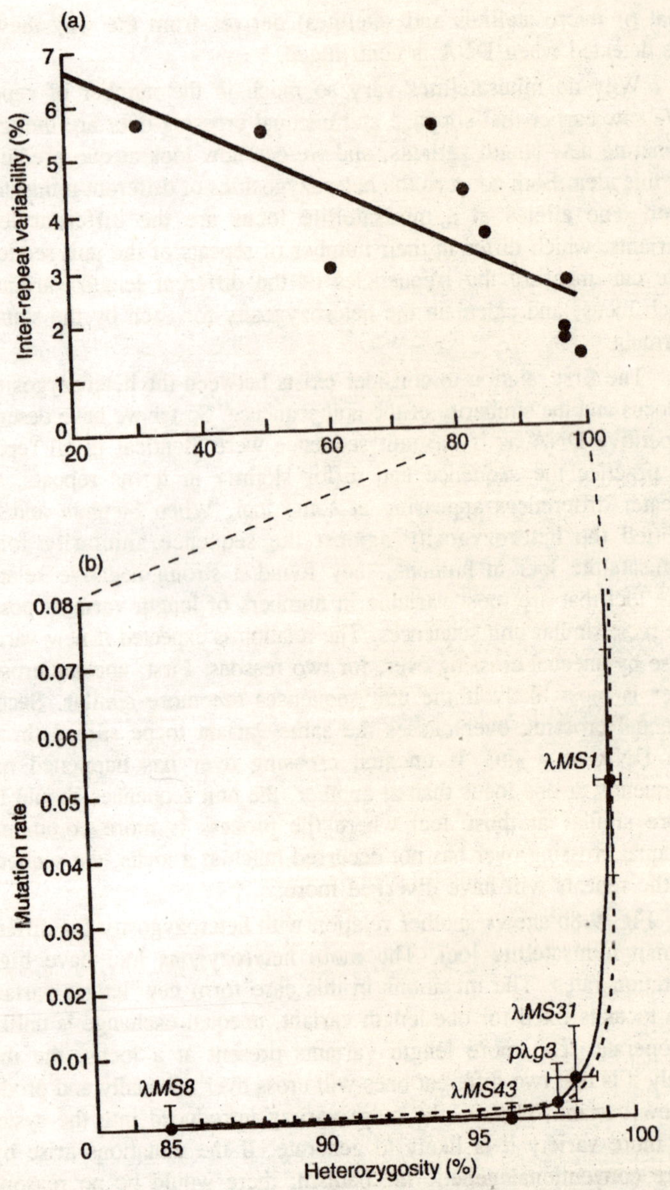

Fig. 9.8. Two relations of the variation of different minisatellite loci in humans. (a) The sequences of the units are more similar at loci that are more heterozygous. (b) Mutation rates are higher at more variable loci. Both relationships make sense if length variants at minisatellites mainly originate by unequal crossing over.

The high mutation rates of minisatellites allow them to be used in *genetic fingerprinting*. New variants arise at a high enough rate for every individual (except monozygous twins) to have a unique frequency distribution (or "profile") of minisatellites. Genetic fingerprints are more forensically useful than real fingerprints. Not only is an individual's genetic fingerprint unique and identifiable from small quantities of body substances, but it is also heritable. The profiles of a father and his children, for instance, are more similar than two random members of the same population. Genetic fingerprinting can, therefore, be used in paternity testing as well as in straightforward identification. Minisatellites similar to those in humans have been found in many other species, and the genetic fingerprinting probe has become an important method of tracing paternity in behavioural ecology. Microsatellites are also used in genetic fingerprinting.

Scattered Repeats

Scattered repetitive DNA is thought to originate mainly by *transposition*. Certain sequences of DNA can, under appropriate circumstances, copy themselves elsewhere in the genome. These mobile genetic sequences are called transposable elements, or (more informally) jumping genes. The molecular biology is complex, but we can distinguish two types of transposition; these processes differ at a molecular level according to whether an RNA intermediate is present. The difference is evolutionarily interesting because transposition by an RNA intermediate may be more likely to increase the number of copies of the sequence in the DNA.

We can make a second distinction between *replicative* and *conservative* transposition. In Fig. 9.9a, two copies of the sequence exist after the transposition, whereas only one was found before transposition. In Fig. 9.9b, however, only one copy exists both before and after transposition. (It is possible for transposition by DNA intermediate to be replicative. For instance, the excised sequence may jump from one of the two strands produced after the DNA has replicated itself in the cell cycle, to a position upstream of the replication fork, where the DNA has yet to be replicated.)

Replicative transposition could theoretically produce a scattered repeat sequence. After its origin in one individual, it could then increase in frequency (by selection or drift) to result in the pattern of scattered repeats we now observe. Did real scattered sequences evolve in this way? The best evidence that a scattered repeat sequence had originated by transposition would be found by observing the sequence transposing.

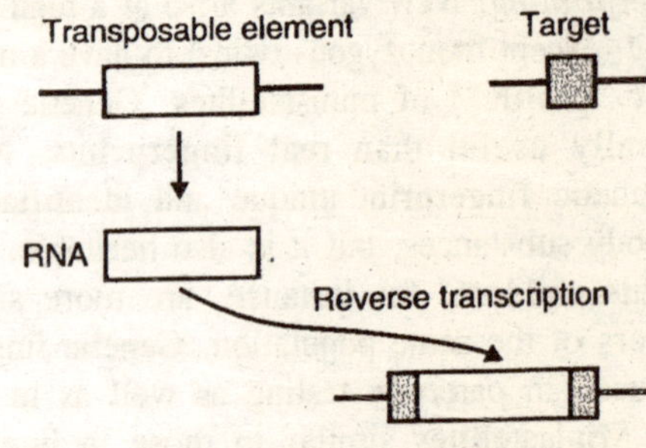

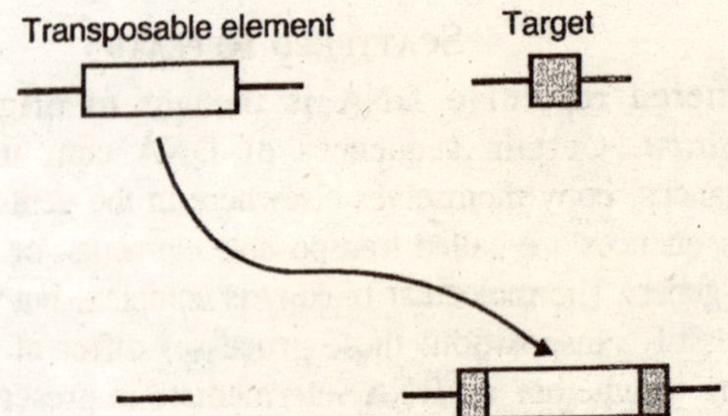

Fig. 9.9. Two main kinds of transposable elements. (a) Retroelements, which transpose by the reverse transcription of an RNA intermediate. (b) Transposable elements, which transpose by excising from one site and inserting elsewhere, without an DNA intermediate.

A more general method is to study the characteristics of the DNA sequences that are known to transpose, and of the scattered repeats, to see whether the latter resemble the former. Transposable elements have some characteristics features, and the similarity between these and at least some well-studied scattered repeat sequences is enough to make a compelling case for their origin by transposition. Let us look first at some known (or strongly suspected) transposable elements, and then at an example of a scattered repeat.

The simplest transposable elements lack the enzyme reverse transcriptase. They possess only a "transposase" enzyme together with one or more other genes, and are transposed by a DNA intermediate. Such elements usually have inverted repeat sequences at either end; the same sequence is found at both ends, with one sequence forming a mirror image of the other. The most famous examples are the first transposable elements to have been described—the ones in maize discovered by McClintock. Other transposable elements have the gene for reverse transcriptase as well, and they are classified as retroelements. Some retroelements can live only in the genome; the element called *copia* in the fruitfly *Drosophila* is an example of this phenomenon. Other retroelements can live independently or within a host genome; these elements include the retroviruses, such as HIV, the agent of AIDS.

Which sequences in our DNA may have been produced from these kinds of transposable elements? Transposition has been most studied in bacteria, and our knowledge for multicellular organisms is less certain, but some likely examples have been identified. The *Alu* sequence is a hot candidate in mammals. *Alu* is the name for a characteristic sequence recognized by the restriction endonuclease *Alu* 1. The sequence is about 281 bases long and is found throughout the human genome; each genome contains about half a million copies. About 5% of human DNA consists of the *Alu* sequence. Few, if any, of the sequences are transcribed, and the *Alu* sequence may have no function.

Where does the *Alu* sequence originate? One favoured hypothesis is that it ultimately comes from the gene for a functional ribosomal RNA molecule called 7SL RNA. The gene for the 7SL RNA could have first been picked up by a retroelement and copied back into the DNA by reverse transcription; the many *Alu* sequences are believed to have been formed as second-order derivatives of the 7SL RNA gene, by recurrent rounds of retroelemental reverse transcription from a number of reverse-transcribed copies of the gene itself. The *Alu* sequence itself is not strictly a retroelement because it lacks the gene for reverse transcription, but it looks like a sequence that is derived from reverse transcription. Perhaps it is picked up by a retroelement, and the gene for reverse transcription is lost when it is copied back into the DNA. Although the sequence has not been proved to be functionless, a strong suspicion exists that it is an example of selfish DNA. The large numbers of copies of *Alu* suggest that something about its sequence makes *Alu* particularly likely to be copied; it would then be an example of active selfish DNA.

Transposable elements can have interesting evolutionary consequences, even if they evolved originally by neutral drift. The insertion of an *Alu* sequence can prevent gene conversion between a pair of genes. Concerted evolution between the two genes then becomes impossible. As a consequence, duplicated genes in which an *Alu* sequence

In terms of 'gene regulation' *per se*, current fashion in recombinant DNA studies focuses heavily on the upstream and downstream landscapes surrounding genes in an effort to elucidate the important sequences used in the control of gene transcription. The existence of various chromatin configurations, enhancers, hypersensitive sites, differential processing at the RNA and protein levels and mRNA stability is relevant only to when, where, and how rapidly, certain proteins are made, modified and dispensed with. They tell us little about developmental regulation, much of which involves interaction between active sites or structural components of proteins and cell-cell interactions. Such interactions are only now being examined by the newer technologies. We focus on them because we believe that it is only by considering cellular phenomena that we will be able to decide which gene pathways are initially worthwhile pursuing in a developmental context.

Cellular Interactions

At gastrulation, entire sheets of cells undergo movement resulting in the particular disposition of the germ layers which allows correct organogenesis to subsequently proceed. What gene programmes underpin these cell determinations and cell movements? If we had some inkling of the genetic pathways involved, we could rationalize how changes in these processes at the cell and tissue level lead to morphological variation. We could then estimate which parts of the pathways were more subject to genetic and environmental stress and hence discover what perturbations are likely to cause significant phenotypic changes during evolution.

Oster and Alberch (1982) provide a striking illustration of a developmental system, epithelial morphogenesis, which brings home the importance of cell-cell interactions and their effects on morphology. The sequence of events that culminates in the formation of feathers, scales, teeth and hair depends initially on cell shape changes involving the epidermis. Here a basal lamina separates an epidermal from a dermal cell layer. The epidermal layer thickens to yield epidermal placodes with cell shape changing from cuboidal to columnar, and cells within the dermal layer then condense beneath the placodes. A binary decision is now made; if *invagination* of the epidermal layer occurs, then either scales (reptiles) or feathers (birds) are formed; if *invagination* is the order of the day, hair or skin glands results (mammals). The critical bifurcation in this system is the inward or outward folding of the epithelial layer. Some relevant questions to be asked in dissecting such a system mechanistically are thus; hcw do we determine the biochemical events that underlie this process? Is it possible to isolate key genes or key proteins which, in the first instance, allow the determinative events to occur?

So shockingly rudimentary is our knowledge of the control of developmental processes that we are forced to start at any chink in the biological armour. Without any doubt, the most promising avenue is the molecular analysis of development in *Drosophila*. Here the application of recombinant DNA and monoclonal technologies, classical genetics, embryology and gene transfer techniques has already revealed that a number of developmentally significant executive genes have a common 61 amino-acid protein domain, termed the homeo box, which is also found in most vertebrates. It appears that the most profitable approach is to continue to isolate, characterize and perturb *Drosophila* systems and then, by raiding vertebrate genomic libraries, to evaluate if the genes and developmental processes are similar or divergent.

Developmental Genes in Drosophila

Despite 30 years of research, the molecular bases of pattern formation are still obscure. We still do not know how the position-dependent activity of molecules is laid down and interpreted. The situation is now rapidly changing owing to the isolation of three major classes of genes which control spatial organization in the *Drosophila* embryo. There are: (1) maternal effect genes that control spatial co-ordinates; (2) segmentation genes that determine body segment number and polarity by interpreting the positional information supplied by the maternal effect genes; and (3) homoeotic genes that specify segment identity. Not only are these genes being cloned and characterized, but in the case of both the sequentation genes and the homoeotics their sites of transcription within the early embryo are being defined by the use of suitable molecular probes.

Nusslein-Volhard and Wieschaus (1980) undertook a saturation coverage of the genome to search for embryonic lethals which altered the segmentation pattern of the embryo. They uncovered more than a dozen loci which were implicated in various stages of segmentation but none of which affected the overall polarity of the embryo. There were genes that effected pattern duplication in each segment, pattern deletion in alternating segments and deletion of adjacent segments.

Other genes have been isolated in less complete searches. One gene, *fushi tarazu* (lack of enough segments), is of particular interest. In homozygous mutant *fushi tarazu* embryos alternate body segments neuronal *connectivity* that has been largely responsible for evolutionary changes in the nervous system. It is all well and good to evolve sight, sound, taste, smell and touch and a neuromuscular system as well as mechanisms for monitoring and co-ordinating internal activities, but these systems need to be integrated in multimodal convergence to give rise to sensible behaviour, and such integration is a function of neurons. Since each human neuron synapses on average with 1000 others, the total connections approach 10^{15}. How are these connections laid down faithfully in any nervous system? What genes or gene families control these processes and how are they altered during evolution to yield novel behaviours?

Once again it will be easier to glean rudimentary principles from invertebrates, but should the reader find this a somewhat presumptuous exercise, we emphasize that nearly half of the monoclonal antibodies directed against *Drosophila* nervous tissue cross-hybridize to the human brain.

Studies using the simple roundworm *Caenorhabditis elegans* make it clear that the development of the nervous system operates through binary decision-making processes. Consider, for example, the fate of six ectoblast cells of the lateral ectoderm of this organism. Five of them undergo a common developmental lineage giving rise to hypodermal cells of the cell surface. One cell, V5, produces both hypodermal and neural cells. This lineage also includes programmed cell death as well as producing a lateral sensory structure known as a postdeirid. This structure consists of a dopaminergic neuron, a non-dopaminergic neuron, a sheath cell and a socket cell.

The homoeotic mutation *lin*-22 disrupts this lineage in a specific way. Cells VI, 2, 3 and 4 now express the exact characteristics of V5, so the organism finds itself with superfluous sensory structures and reciprocally lacks hypodermal cells.

There is also a gene, *ced*-3, the mutant form of which rescues cells that normally are destined to undergo programmed cell death. In the V5 lineage described above, a non-dopaminergic neuron and cell death was the normal outcome. In the presence of *ced*-3, however, death does not occur and in- stead a supernumerary dopaminergic neuron is generated.

Caenorhabditis is also in the vanguard for the analysis of heterochronic genes. Changes in the temporal pattern of gene expression, and their concomitant morphological novelties has always ranked high as prime movers of evolution. In this organism, two major classes of mutants can be summarized as follows: in *delayed* (or retarded) cases juvenile characteristics are expressed in the adult, whereas in *early* (or precocious) cases adult characteristics appear in immature individuals. Ambros and Horvitz (1984) have isolated a number of such mutants. In *lin*-14, -28 and -29, for example, certain developmental events occur either earlier or later than normal. Particular *lin*-14 alleles can lead to reiterations where moulting and larval cuticle events are repeated. In other alleles, a hypodermal blast cell does not produce its normal progeny during the L2 stage but generates a lineage produced a larval generation earlier by its grandparental precursor cell.

The important point to note from this simple organism is that genes can be identified which cause heterochronic changes. Their relevance in an evolutionary context will be discernable when they are cloned and their protein products analyzed, and any cross homologies to other organisms evaluated.

Neural Pathfinding Genes and Phenotype

When one examines the setting up of a nerve network, the problems, the genes and the final phenotype, are far from simple. Once again the initial inroads stem from molecular approaches in invertebrates since here the developing neurons consist of a segmental chain which arises from the metameric segmentation of the arthropod body. The studies of Thomas et al. (1984) and Goodman et al. (1984) reveal that in the grasshopper the early growth cones of nerve cells send out filopodia and use cell surface cues to build an initial scaffold along which later developing neurons find their way. Thus a developing G neuron, for example, makes a choice when it encounters the A/P nerve bundle. Which of the four neurons within this bundle will it follow? It makes its choice on the basis of cell-surface molecules on axons and their selective adhesion properties.

Rather surprisingly, when other arthropods are examined, such as a crayfish (*Procambarus*) and *Drosophila*, astounding similarities are seen in the location of identified neurons, in the location of cell bodies, in the axonal scaffolds and in the bifurcations that are made by the respective growth cones. For example, the G neuron in *Drosophila* has near it at least 25 nerve bundles and yet it invariably selects the A/P bundle and within it the P, but not the A, axons, just as does the grasshopper.

Thus, in examining the descriptive biology of how the embryonic nervous system develops, a remarkable early conservatism is seen in arthropods as different as crayfish, grasshopper and *Drosophila*. The important future steps will be first, to describe the network and discover the embryonic rules to see how they may be perturbed; secondly, to determine what causes different neurons to express specific surface labels; thirdly, to determine how growth cones are guided to their final destinations on the basis of selective substrate affinities and fourthly, to determine the molecular basis of cell recognition during neuronal development.

In order to answer some of these questions, one needs to isolate the molecules involved or the genes that code for them. The use of monoclonal antibodies has already revealed a specificity that is not apparent on morphological inspection. In the grasshopper, the monoclonal antibody *Mes*-2 recognizes an antigen that is present on only four neurons out of 1000 in a hemisegment. *Mes*-3 and *Mes*-4, on the other hand, distinguish the MP1-dMP2 nerve bundle from the other 25 longitudinal fascicles that occur at this developmental stage.

These findings mean that we are on the threshold of unravelling simple circuits in terms of cell-surface molecules. What is still unknown, is whether these phenomena are controlled by single genes or multigene families and whether there are many different proteins, or whether the proteins are few but their modification, by glycosylation, leads to a diverse array of cell-surface types.

Since the wiring diagram of the central nervous systems of both vertebrates and invertebrates is so poorly described, it is difficult to take the alternative genetic approach of mutagenizing the genome and looking for specific connectivity dislocations in anything but giant fibres or anatomically distinct cell types. This approach, however, has also begun in *Drosophila*.

In the vertebrate brain the molecular situation is almost palaeoneuro-biological. Even so, the surprises are already with us. One will recall that, relative to other tissues, the polyA$^+$ RNA complexity of the vertebrate brain is enormous, near 120 million nucleotides, whereas that of liver and kidney, for example, is of the order of 30 million nucleotides. Milner and Sutcliffe (1983) and Sutcliffe et al. (1983) have isolated polyA$^+$ RNA from adult rat brain, converted it to cDNA and have found that brain specific RNAs are larger and rarer than most metabolic transcripts. They estimate that there are at least 30,000 polyA$^+$ RNAs in the brain, most of which are brain specific. This is a larger number of transcripts than all other somatic tissues combined. Furthermore, brain-specific messages contain an 82 bp core sequence, termed an identifier (ID). There are about 100,000 of these sequences per genome and they are generally found in the introns of brain genes where they have been hypothesized to be important in brain-specific gene control.

Sutcliffe et al. (1983) also determined the nucleotide sequence of one of their clones (IB236), identified the sense strand and its long open reading frame and translated it into the amino acid sequence. They then chemically synthesized short peptides, raised antisera against such molecules, and then probed thin brain sections in immunocytochemical reactions. These experiments revealed a neuronal specificity which was not apparent from neuronal morphology alone. Clone IB236 is synthesized in certain cell bodies and transported via certain axons to particular areas of the cerebellum, hippocampus and cortex.

These experiments, although only in the pioneering stage, lift the lid off the problems, complexities and evolutionary hurdles that need to be faced in the central nervous systems of vertebrates and invertebrates. Some of the structural components will clearly be held in common, as we saw from the *Drosophila* monoclonal cross-hybridization to human brain, but others will clearly be distinctive. Furthermore, if brain-specific transcripts, or the protein products, really are larger than conventional metabolic transcripts, then the developmental pathways so involved may have a greater complexity. It is already known that several neuropeptide precursors such as pro-piomelanocortin are processed to yield several proteins.

All these results, both anatomical and molecular, are still too patchy to give us a firm idea of the underlying principles involved in setting up the architecture of the brain. However, the pioneering approaches we have illustrated show that the gap between genes and phenotype is rapidly narrowing.

Genes, Phenotype and Evolutionary Studies

When one examines the controversies that still rage in population genetics, namely selectionists versus neutralists, gradualists versus punctualists, and regulatory versus structural genes as prime movers of evolutionary novelty, one is struck by how far removed they are from the real problems of evolution. In the case of the structural versus regulatory gene argument, it has quite correctly been pointed out by MacIntyre (1982) that no single compelling observation was responsible for this spotlighting of the putatively significant evolutionary role of regulatory genes. In fact '... in the absence of detailed molecular and developmental studies, any interpretation of interspecific differences involving gene regulation are strictly hypothetical'.

We turn finally from the reductionist molecular approach, which is now pouring out information on genomic architecture, and ask how much of it impinges on evolutionary studies. Here, much of the data are being used to examine rates of change in nucleotide and protein sequences, and to refine phylogenies by the estimations of evolutionary distances. Very little of it addresses itself to the underlying mechanisms of phenotypic change that have given rise to morphological novelties. Molecular evolution is currently still in a DNA data collecting phase, and does not address itself to what many consider to be the main tasks of population genetics, namely '... the description and prediction of changes in the genetic and phenotypic composition of populations responding to the forces of mutation, recombination, selection, migration and systems of mating'.

Multigene Families and Molecular Drive

In general, the mechanisms by which new genes and new phenotypes arise is largely based on speculation. Multigene families were, until recently, not an integral part of modern theories, and as yet they still sit uneasily with Neo-Darwinists.

The concept of molecular drive as developed by Dover (1982), and its implication for evolutionary change as well as phenotypic novelty has exposed the fragility of some aspects of Neo-Darwinism. This concept is based on a feature of many multigene and non-coding families (ribosomal DNA, globins, immunoglobulins, histones and repetitious 'satellite' sequences) in which a continuous homogenization process produces a greater within-species homogeneity than between-species homogeneity.

In utilizing turnover in multigene families, drive focuses on the spread and establishment of novelties within populations by internal processes of flux such as unequal exchange, transposition and gene conversion. This literally leads to the transformation of a population under its own steam without the involvement of any conventional selective forces. The example of the bat-eared fox used by Dover (1984), illustrates the situation, though it could be applied to any presumed adaptation.

Bat-eared foxes have abnormally large bat-like ears and the conventional 'explanation' for such extremities is that they have evolved under the aegis of natural selection for increased efficiency in tracking prey. They would conventionally be considered as evidence of an adaptation that increases the survival of the animals possessing them, as most adaptations are so interpreted. There are alternative explanations, however, particularly if multigene families are involved, and we consider them below.

What Dover's theory encapsulates is that once the nexus is broken between chromosome behaviour and gene behaviour, as occurs by internal turnover mechanisms in the genome, then there can be unexpected transformations of a population with the changes occurring in a cohesive manner. Different individuals in a population undergoing molecular drive will be expected to show little variance between members of the particular multigene family in question. Individual members of a multigene family are not constrained by a chromosome, since chromosomes no longer provide refugia from the genomic turnover processes, which can be both inter- and intrachromosomal. Thus when the rate of spread of a variant between chromosomes differs from the rate of gametic mixing, a multigene family can be homogenized in a synchronous fashion in all individuals. This means that a *population* is slowly changing from one DNA composition to another *without* a large variance between individuals. Thus if the bat-like ears of the fox are largely determined by one or a number of multigene families (and this is as yet unknown), then the change to large ears may well be a consequence of genomic turnover processes. This increase in ear size affects all members of the population because it is a *cohesive* molecular change. The change in phenotype can be gradual or sudden, and there are no biological incompatibilities between individuals.

This last point circumvents the problem of how a *single* mutant individual arising in a population can lead to subsequent fixation of a new phenotype. Under molecular drive, the DNA of all individuals can be converted slowly and synchronously. Thus the change of ear size in foxes could simply be a fortuitous by-product of changes in multigene families that impose no real problems of differential survival. Nor do such changes need to be adaptive.

We see from examples such as these that much of the fabric of population genetics theory is under pressure to accommodate the molecular data bases. Furthermore, it is particularly striking that the field not only shows strong resistance to any challenge of orthodox beliefs, but is itself preoccupied with what Levin (1984) has referred to as the 'two camp syndrome':

> Some hypotheses are more often championed than tested. Positions (almost two for any given issue) are fiercely defended even when they are not mutually exclusive. New ideas are liable to attack for no reasons other than their real, or even apparent, violation of orthodoxy.

It is clear to us that this preoccupation with Darwinian orthodoxy enshrouds the field to little benefit. Dover's concept of molecular drive is a bold new incursion into the orthodox realm. It is based on known mechanisms of genomic turnover processes, whereas selection has always been, and still is, more a logical construct than a well-documented process. Neo-Darwinism as an 'explanation' for evolutionary change is in fact in need of a serious new outlook. It needs to focus back to the developmental gene circuits that are critical for phenotypic change and hence for evolutionary change.

INDEX